Sir Edward Albert Sharpey-Schafer

A Course of Practical Histology

Sir Edward Albert Sharpey-Schafer

A Course of Practical Histology

ISBN/EAN: 9783337399092

Printed in Europe, USA, Canada, Australia, Japan

Cover: Foto ©berggeist007 / pixelio.de

More available books at **www.hansebooks.com**

A COURSE

OF

PRACTICAL HISTOLOGY

BY

EDWARD ALBERT SCHÄFER, LL.D., F.R.S.

PHILADELPHIA
LEA BROTHERS & CO.
1897

PREFACE

THE purpose of this work is to afford to those engaged in the practical study of Histology plain and intelligible directions for the suitable preparation of the animal tissues, with the object either of immediate study or of their preservation as specimens for future reference. The aim throughout has been to assist the student to carry on histological work independently of the constant presence of a teacher. Care has therefore been taken to select methods upon which, in the author's experience, complete reliance can be placed — provided, of course, that the directions given are faithfully followed ; and no attempt has been made to give anything approaching a complete account of modern methods, the multiplicity of which is bewildering, and many of which have been only devised for a particular line of research. Moreover, such an account would have been unnecessary, since (to the great convenience of investigators) it is already furnished by Dr. A. Bolles Lee in his book 'The Microtomist's Vade Mecum.'

In an Introductory Chapter an account is given of the several parts of the microscope, and the purpose for which they are intended (without entering into an explanation of its optical construction), and of the instruments and the methods of staining and otherwise preparing tissues which are in more general use in Histology. Here also instructions will be found for measuring and for delineating microscopic objects. The more special methods of preparation are given with the tissues and organs for which they are employed.

Throughout the book detailed descriptions of tissues have been avoided, seeing that these are to be found in systematic works, such as Quain's 'Anatomy,' Klein's 'Elements of Histology,' or the author's 'Essentials of Histology.'

UNIVERSITY COLLEGE, LONDON:
   May 1897.

CONTENTS

CHAPTER		PAGE
	INTRODUCTORY: INSTRUMENTS AND GENERAL METHODS USED IN HISTOLOGY	1
I.	THE BLOOD	48
II.	THE EPITHELIAL TISSUES	82
III.	CONNECTIVE TISSUE	96
IV.	CARTILAGE	114
V.	BONE	123
VI.	MUSCULAR TISSUE	134
VII.	NERVOUS TISSUE	146
VIII.	THE BLOOD-VESSELS	162
IX.	LYMPHATICS AND SEROUS MEMBRANES. SYNOVIAL MEMBRANES	186
X.	THE SKIN, HAIRS, AND NAILS	199
XI.	THE HEART	205
XII.	THE LUNGS	208
XIII.	THE MOUTH AND PHARYNX. TEETH. TONGUE. SALIVARY GLANDS	213
XIV.	THE ŒSOPHAGUS AND STOMACH	220
XV.	THE SMALL AND LARGE INTESTINES	225
XVI.	THE LIVER AND PANCREAS	231
XVII.	THE DUCTLESS GLANDS	236

CHAPTER		PAGE
XVIII.	THE URINARY ORGANS	242
XIX.	THE GENERATIVE ORGANS	247
XX.	THE CENTRAL NERVOUS SYSTEM	252
XXI.	THE ORGANS OF THE SENSES—THE EYE	257
XXII.	THE AUDITORY, OLFACTORY, AND GUSTATORY ORGANS	283
	INDEX	291

LIST OF ILLUSTRATIONS

FIG.		PAGE
1.	Diagram of Microscope.	2
2.	Iris Diaphragm.	3
3.	Sub-stage Condenser	3
4.	Simple Microscope	6
5.	Compound Erecting Microscope for Dissecting.	7
6.	Glass Slide and Cover-glass.	8
7.	Forceps for taking up Cover-glass	9
8.	Clip to hold Cover-glass	9
9.	Spear-headed Needle.	9
10.	Scissors, Forceps, and Mounted Needle	10
11.	Section-lifters	12
12.	Mode of using Pipette for Irrigation.	13
13.	Card Tray for holding Microscope Slides	13
14.	Bottle for Xylol Balsam	25
15.	Outline of Paper for Embedding Trough.	26
16.	Paper folded to form Trough	27
17.	Embedding Trough of Lead Foil	27
18.	Section-cutting with the Free Hand.	28
19.	Freezing Microtome (Ether Spray), Jung's	29
20.	Brass Holders for Collodion Preparations.	30

FIG.		PAGE
21.	Inclined Plane Microtome, Jung's	31
22.	L-Shaped Brass Blocks forming Mould for Paraffin	32
23.	Box for storing Paraffin Blocks	33
24.	Tripod Microtome (Birch's Pattern)	34
25.	Glass Pot with Cover	35
26.	Rocking Microtome (Cambridge Pattern)	35
27.	Minot's Microtome	36
28.	Grooved Trough to hold Microscope Slides	38
29.	Camera Lucida, Zeiss	40
30.	Camera Lucida of Abbe	41
31.	Upright Micro-photographic Camera	42
32.	Horizontal Micro-photographic Camera	43
33.	Ocular Micrometer	45
34.	Lines of Stage Micrometer viewed with Ocular Micrometer	46
35.	Simple Warm Stage	53
36.	Mode of using Simple Warm Stage	54
37.	Warm Stage with Gas Regulator	56
38.	Box to inclose Microscope, with Gas Regulator and Thermometer	58
39.	Simple Moist Chamber	62
40.	Gas Chamber	63
41.	Slide ruled in Squares of 0·1 mm. each for Enumeration of Blood-corpuscles	64
42.	Oliver's Apparatus for Estimating the Number of Blood-corpuscles	67
43.	White Corpuscles migrating from Clot	71
44.	Slides for the Passage of Electric Shocks through a Preparation	72
45.	Apparatus for Passing Electric Shocks through a Preparation	73

LIST OF ILLUSTRATIONS xi

FIG.		PAGE
46.	Apparatus for Passing CO_2 over a Preparation	76
47.	Valve of Mussel, showing Gills	90
48.	Apparatus for Passing CO_2	93
49.	Syringe for Interstitial Injection	101
50.	Warming Apparatus with Gas Regulator	119
51.	Frog-Cork for viewing Circulation in Web, Tongue, or Mesentery	171
52.	Structure and Position of Tongue of Toad	174
53.	Injecting Apparatus for Blood-Vessels	180
54.	Cannulas and Clips used in Injecting	182
55.	Hoggan's Rings	189
56.	Mercury-pressure Apparatus for Injecting Lymphatics	193
57.	Fine Perforated Steel Needle for Injecting Lymphatics	194
58.	Mode of cutting Frog's Cornea to cause it to lie flat	264
59.	Tongue of Rabbit	289

PRACTICAL HISTOLOGY

INTRODUCTORY

INSTRUMENTS USED IN HISTOLOGY

The Microscope.—The practical study of Histology is mainly dependent upon the use of the microscope. The microscope is a combination of lenses arranged for the purpose of obtaining and viewing a magnified image of any minute object. The lenses are set in a tube of variable length—*the tube of the microscope* (fig. 1, t, t')—and this is itself supported in a vertical position, on a firm metal *stand*, which is provided with an arrangement by which the tube is capable of being moved, without lateral deviation, in a perfectly straight, up-and-down direction. This arrangement is termed the *adjustment*. Its purpose is to bring the microscope into that position with regard to the object in which the latter is most clearly seen. The object is then said to be in focus.

Two adjustments are commonly provided : one—the *coarse adjustment* (*Adj*)—serves to bring the lenses roughly into the focal position, and is either a telescopic joint or a rack and pinion movement ; the other—the *fine adjustment* (*adj'*)—is a fine screw, and by its means the focus may be obtained with complete exactness even when the highest magnifying powers are employed. The stand is further provided with a rigidly connected, horizontal table or *stage* (*st*), upon which the object is placed, and which projects below the tube, and is provided

B

PRACTICAL HISTOLOGY

Fig. 1

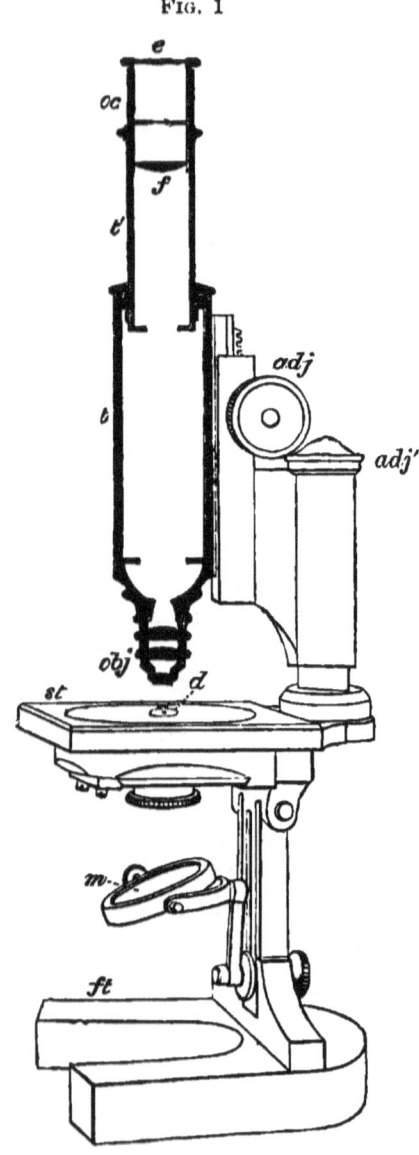

Diagram of microscope

t, tube seen in section ; *t'*, sliding part of tube ; *oc*, ocular ; with *e*, eye-glass, and *f*, field-glass ; *obj*, objective ; *adj*, coarse, and *adj'* fine adjustment ; *st*, stage ; with *d*, diaphragm ; *m*, mirror ; *ft*, foot of microscope

The object is indicated by a small arrow just above the aperture in the diaphragm ; the magnified image by the larger arrow in the middle of the ocular with a circular aperture to admit light from below to the object, capable of being varied by means of an *iris-diaphragm* (fig. 2), or by *stops* (fig. 1, *d*) furnished with holes of different sizes. Diffused daylight is, if possible, employed, and is reflected, by means of a movable *mirror* (*m*) below the stage, upwards through the object and through the tube of the microscope to the eye of the observer. This is termed viewing an object by transmitted light. Occasionally, especially when comparatively large and opaque objects and low magnifying powers are to be employed, the former are viewed by the light which is re-

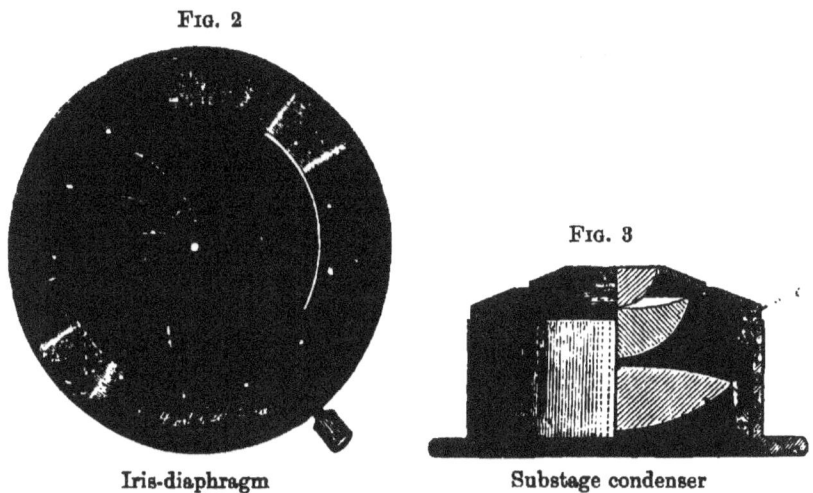

Fig. 2

Fig. 3

Iris-diaphragm　　　　　　Substage condenser

flected from their surface, whilst that from the mirror is cut off. In order in such cases to concentrate as much light as possible upon the object a bull's-eye condenser is employed. It is only in viewing such preparations that the binocular microscope offers any material advantage in histology.

To obtain a more central and concentrated illumination for objects which are to be viewed by transmitted light, a condenser is frequently fitted below the stage of the microscope (*substage condenser*, fig. 3). The stage may conveniently be ruled at one part with numbered lines crossing one another at right angles. By this means the position of a slide can be recorded, and an object readily found again by replacing the slide in that position.

The lenses form the essential part of the microscope, and are so arranged that one combination of them is placed at the lower end of the tube and produces a magnified, inverted image of the object at the upper part. The image thus produced is viewed, and at the same time still further magnified, by another combination at the top of the microscope tube. The lenses at the lower end of the tube form an achromatic combination which, from its situation near the object, is termed the *object-glass* or *objective*, and it is upon the perfection of its construction that the usefulness of the microscope mainly depends. The combination which is at the upper end of the microscope tube is termed the *eye-piece* or *ocular*, and consists of a lens, the *eye-piece proper* (e), placed close to the eye of the observer, and of another glass (the *field-glass*, f) situate below the first and having for its object the collection of the more divergent rays transmitted by the objective, and also the lessening of chromatic aberration. The whole ocular (oc) thus composed is made to slip in at the superior aperture of the microscope tube. It is seldom necessary to have more than one ocular, of medium strength, in use, but at least two objectives of different magnifying powers are essential for ordinary histological work. One of these, which in subsequent pages will be spoken of as the *low power*, should, when used with the ordinary eye-piece, give an apparent linear enlargement of about 75 diameters; that is to say, when a line, the length of which is known (say $\frac{1}{100}$ of an inch), is observed through this combination it should appear seventy-five times as long as it really is ($\frac{3}{4}$ inch therefore). The other, to be mentioned as the *high power*, should given an apparent linear magnification of from 300 to 400 diameters. These two glasses amply suffice for all ordinary histological studies; but for certain special subjects it is advantageous to obtain the use of a more powerful combination—one that will magnify 1,000 diameters or more.

Glasses of this high magnifying power are usually of the kind known as 'immersion-objectives,' so called because they are corrected for use with a stratum of *distilled water* or of *cedar-oil* between the specimen and the lower lens of the objective. It is best to apply the water or oil to this lens with a splinter of wood or from a pipette before the objective is screwed on to the tube of the microscope; the tube is then lowered by means of the coarse adjustment until the drop of liquid comes in contact with the cover-glass, after which the focus is obtained by cautiously lowering it further, at first with the coarse, and then with the fine adjustment.

Apochromatic objectives (with compensating oculars), which were introduced by Abbe and Zeiss, are now furnished by all good microscope makers. They are distinctly superior to the ordinary objectives.

The above comprise the essential requirements of a microscope, but larger instruments are often provided with certain adjuncts which render them in a measure more complete. The stand is usually hinged, so that the stage and tube can be tilted somewhat out of the perpendicular to allow of better adaptation to the position of the observer. But if the microscope tube is not inconveniently high, it is almost as comfortable to work without inclining the instrument, and some preparations, those, for instance, which have to be examined in fluid, will not admit of inclination. A *mechanical stage*, i.e. a stage capable of carrying the object under examination horizontally in two directions by screw adjustment or other mechanical means, instead of by hand, is sometimes added, but for most work can very well be dispensed with. A *camera lucida* (see p. 41) is useful for obtaining an exact sketch of the outlines of an object. A *nose-piece*, holding two or even three or more objectives, which can be changed by a simple act of rotation, is a very convenient adjunct.

A *polarisation apparatus* is occasionally employed in investigating the optical properties of the substances which compose the tissues, and is also of use in helping to determine the nature of crystalline deposits in urine and other fluids.

In connection with the employment of this the stage-plate of the microscope is, in the best instruments, made capable of rotating on a vertical axis. In smaller instruments this movement is, as a rule, not provided for, and, indeed, although convenient, is not essential.[1]

Perfect steadiness of the stand, stage, and adjustments is the most important point to pay attention to in selecting a

FIG. 4

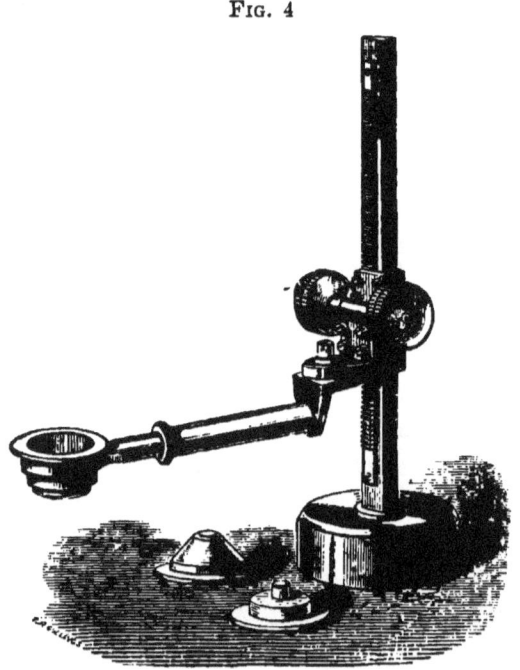

Simple form of dissecting microscope, provided with three lenses of different magnifying power

useful microscope, so far as the body of the instrument is concerned. The excellence of the objectives can only be competently judged of by one who is already somewhat conversant with the use of the microscope.

In addition to the ordinary instrument (which is generally

---

[1] A description of the method of using the polarisation apparatus is given in the chapter on muscle (p. 142).

distinguished as the 'compound' microscope), it will be found very convenient to have a simple microscope as well : one is thus better able to follow the needles or other instruments when engaged in the separation and manipulation of minute objects. An instrument which is used for this purpose is termed a *dissecting* or *preparing microscope*. Any simple lens which is mounted on a stand will serve (fig. 4), and even the bull's-eye condenser, which is frequently furnished with the microscope, may be employed as a lens if the marginal part is

FIG. 5

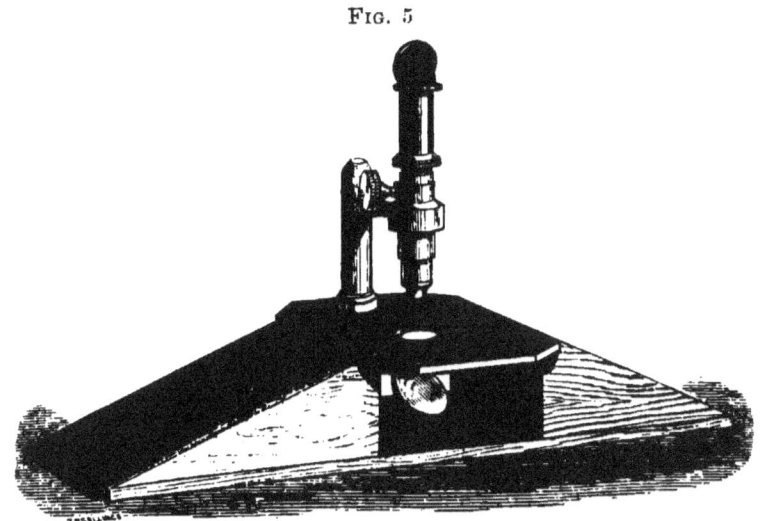

Compound dissecting microscope, made by Nachet

covered by a black-paper diaphragm. A very convenient form of dissecting microscope is shown in fig. 5. This consists of a small compound microscope with a low-power objective and furnished with an ocular constructed for reversing the image, so that the object appears in its natural position and not inverted. It is provided also with a prism for reflecting the light in a direction convenient for the eye, and is placed on a wooden stand, so constructed as to afford support to the arms of the operator.

Instruments.—Besides the microscope, the student who is commencing the practical study of histology will find it necessary to be provided with the following simple instruments and appliances :—

Glass-slides and cover-glasses (see fig. 6).—Microscopic slides are oblong slips of glass upon which the object is placed. In this country they are usually cut of the uniform and convenient size of three inches by one inch. The glass should be quite free from flaws and specks. The cover-glasses should be the finest sold. It is true that when very delicate they are likely to be broken in the cleaning unless great care is exercised ; but, on the other hand, some high-power objectives require to be focussed so close to the object that a thick cover-

FIG. 6

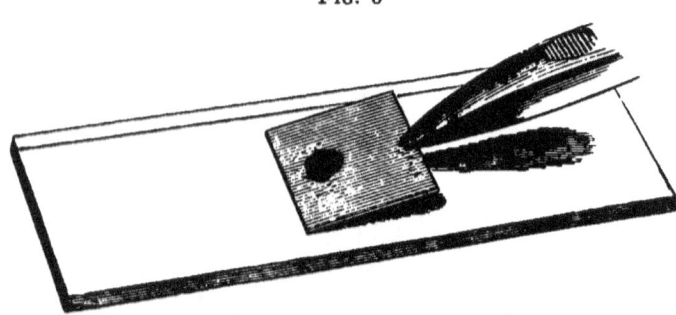

Glass slide and cover-glass, natural size

The figure shows the mode of letting down the cover-glass (with a drop of fluid on its under surface) gently on to the middle of the slide

glass cannot be used without risk of crushing the tissue, and perhaps of scratching the objective. As a rule, the chief difficulty in cleaning cover-glasses arises from the fact that their surfaces are often covered with a thin film of grease or other organic matter which it is almost impossible to rub off. But this is easily got rid of by placing a number of them together in a small glass beaker and pouring a little strong nitric acid upon them : this quickly destroys every trace of organic matter. The acid is then poured off, and the cover-glasses are thoroughly rinsed by allowing water to stream over them from a tap for two or three minutes. They may then be kept in distilled water or spirit ready for use, and need only be dried when wanted. The drying is effected by a thin linen or cotton cloth, which is laid flat on the table, and the cover-glass, having been placed on this, is gently rubbed, first on the one surface and then on the other, with a corner of the cloth. Either ordinary or special forceps (fig. 7) may be used for placing the cover-glass in position

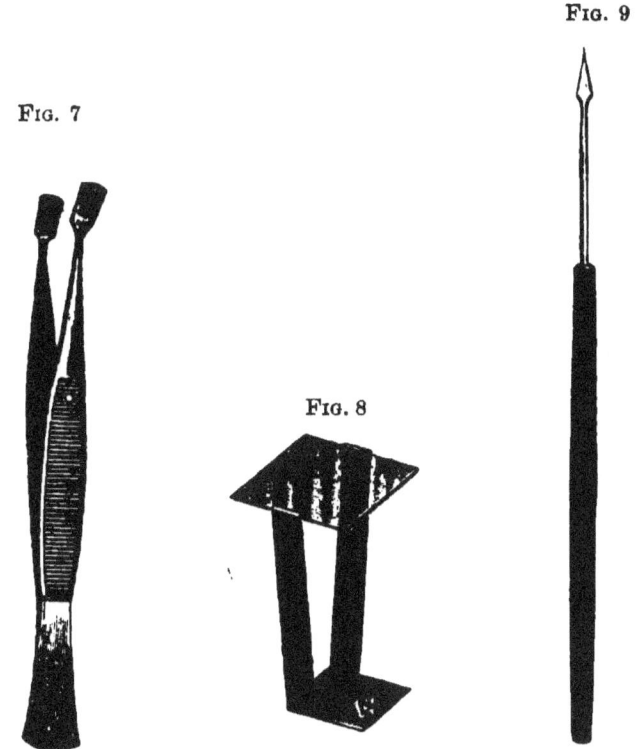

FIG. 7

FIG. 8

FIG. 9

Cover-glass forceps   Cover-glass holder (Zoth)   Spear-headed needle

on the slide, but, as a rule, an expert microscopist can perform the operation equally well with the cover-glass held between the thumb and forefinger applied to opposite edges. When cleaned and dried ready for use the cover-glass may be held in a small metallic clip (fig. 8). Square cover-glasses should always be used in preference to round ones when reagents have to be applied to a specimen under the microscope. The

Fig. 10

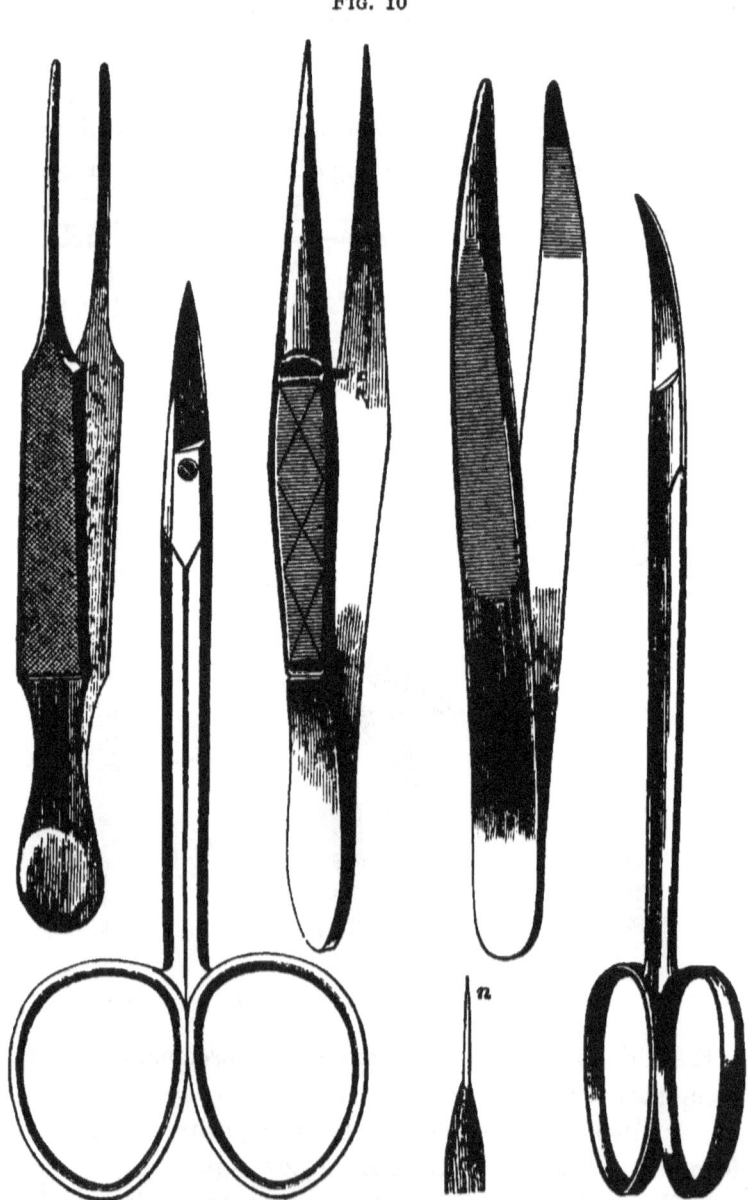

Scissors and forceps suitable for histological purposes. Natural size

n, end of a mounted needle

most convenient size for general purposes is $\frac{3}{4}$ inch (about 2 centimetres) square.

Mounted needles.—These are sewing-needles mounted in a wooden handle, with about $\frac{5}{8}$ inch (1·5 centimetres) of their pointed end projecting (fig. 10, *n*). They are amongst the most useful instruments which the histologist possesses, and will be in constant requisition. They must always be kept clean and sharp. A spear-headed needle is also frequently a useful adjunct (fig. 9).

Scissors and forceps.—A small pair of scissors with short, straight blades is necessary. Their cutting edge must always be kept very sharp, especially towards the point. A small curved pair is often useful. At least two pairs of small steel forceps are requisite. The blades should be short, broad at their junction, so as not to admit of lateral deviation, and tapering to a blunt point. They are slightly roughened at the end, so as to afford a firmer grasp.

Section-lifters.—One or two section-lifters (fig. 11) of different sizes are required. They are made of platinum, steel, copper, or German silver, and may be either flat or slightly concave, in the latter case with a small hole in the centre.

Slips of white blotting paper should always be at hand. They serve both for soaking up excess of fluid from under the cover-glass and for placing the slide upon, when preparing a tissue that has been stained, so that it is better seen than it would be upon a black surface. On the other hand, it is better to use a black surface for working upon when tissues are unstained. A number of *shallow glass or porcelain pots* provided with well-fitting covers (see fig. 25), and one or two large *shallow glass or porcelain dishes* (such as are used by photographers) are in constant requisition, since very many of the manipulations require to be performed whilst the tissue is immersed and floated out in fluid ; while for applying reagents to a specimen under the microscope small *pipettes* (fig. 12) are extremely convenient. Pipettes may be readily made with the aid of a blow-pipe by drawing out a piece of soft

12 PRACTICAL HISTOLOGY

Fig. 11

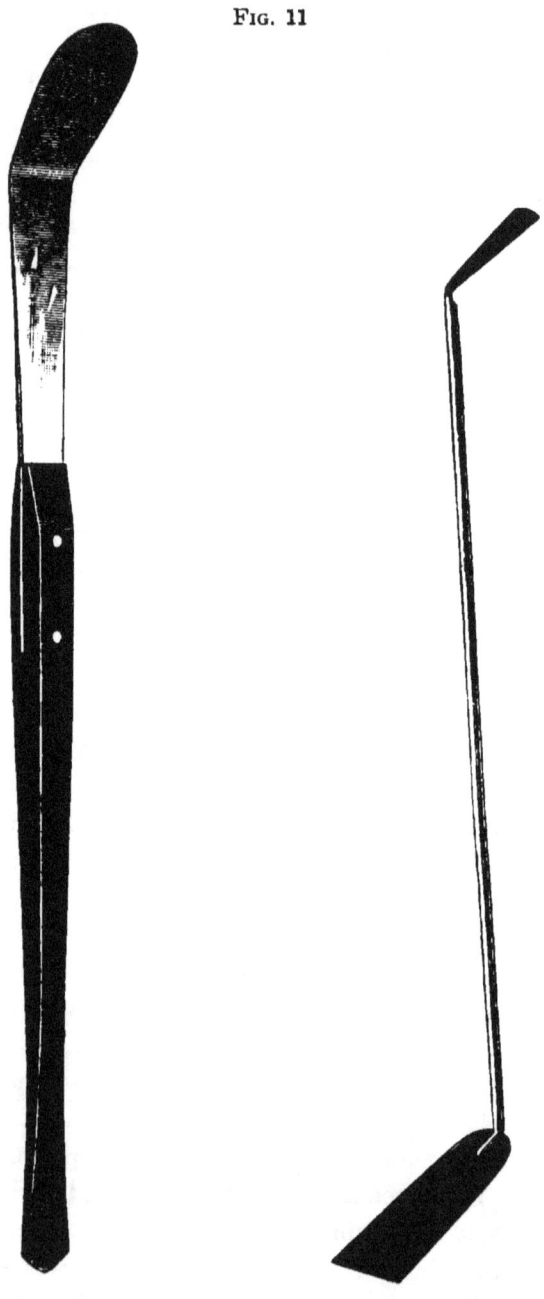

Section-lifters. Natural size

INSTRUMENTS

Fig. 12

Method of irrigating a preparation with fluid from a capillary pipette
p, blotting paper

Fig. 13

Card tray for holding twenty microscope slides

glass tube at two places close to one another, so that the intermediate part remains as the bulb of the pipette. It is well to make a number at a time, sealing up the ends in the flame while the bulb is still hot; they are thus sterilised and made dust-tight, and may be kept always ready for use. It is only necessary to break off the sealed ends when required and to suck the re-agent up into the bulb. A pipette should always be rejected after it has once been used, and never employed for another reagent.

For keeping preparations that have been permanently mounted, flat trays to hold twenty or more specimens are the most useful (fig. 13). They are conveniently made of cardboard or *papier mâché* with the sides folding over the middle, so that the preparations are completely shielded from dust, and may thus be arranged on shelves or in cabinets. Boxes with grooved sides in which the specimens stand vertically are also very convenient and of less cost.

REAGENTS USED IN HISTOLOGY

A vast number of reagents are from time to time and for various purposes employed in special histological work, but comparatively few are needed for ordinary work. These, therefore, will alone be mentioned in this place, others being referred to as they are required for any particular object.

SOLUTIONS FOR THE EXAMINATION OF THE FRESH TISSUES

Serum of blood and lymph form the most natural fluids in which to prepare fresh tissues for examination. As a form of lymph, the aqueous humour of the eye may be also employed, and has the advantage that it can always easily be got from a recently killed animal. But the albuminous nature of these fluids is to this extent an objection to their use, that it prevents the ready employment of hardening or fixing reagents if it should be necessary afterwards to use these. This objection does not apply to salt solution, which in the form of the so-called **normal saline solution** (common salt 6 to 9 parts, tap-water 1,000 parts) is extensively used. It must be understood that a strength of salt solution which is normal for one animal may not be so for another. Thus a saline solution of 6 parts to 1,000, which is normal or isotonic for frogs' blood-corpuscles (*i.e.* tends to produce no alteration in their form by either diffusing into or abstracting water from them), is subnormal (hypotonic) for mammalian corpuscles, which require 9 parts per 1,000 to give a like result.

SOLUTIONS FOR FIXING THE TISSUES AND ORGANS WITH THE FORM AND STRUCTURE THEY POSSESS DURING LIFE

Many of the most valuable solutions for this purpose are of an acid nature. Those most commonly employed are *chromic acid* and the *bichromates, osmic acid, acetic acid, picric acid, nitric acid, corrosive sublimate*, or mixtures of two or more of these; last but not least, *alcohol*. Recently also *formaldehyde* has been introduced for the same purpose. The strength of the solutions used varies with the purpose for which they are to be employed. If the elements of the tissue are to be dissociated, weak solutions of the fixing fluids are taken; if its parts are to be kept together and the pieces hardened for the preparation of sections, stronger solutions are employed. In every case the fluid should far exceed in bulk the piece of tissue or organ to be preserved, and the latter should never be too thick for the preserving fluid readily to penetrate to every part. It is best, if possible, to inject it into the blood-vessels or into the interstices of a tissue.

Chromic acid.—This may be kept as a 1 per cent. solution, to be diluted as required. Chromic acid is an invaluable reagent for fixing and hardening, and for this purpose a strength of solution varying from 0·2 per cent. to 1 per cent. may be used. It is best, as a rule, to use the weaker solutions, since the stronger are liable to interfere with the staining of the preparations by dyes. Tissues are hardened by chromic acid in a few days. For use as a dissociant the solution may be diluted down to 0·05 per cent. ($\frac{1}{20}$ per cent.) or more.

Bichromates.—Bichromate of potash and bichromate of ammonia are both used extensively in histology, in solutions of 2 per cent. to 3 per cent. They have the advantage over chromic acid that they penetrate a piece of tissue more rapidly, so that larger portions of organs can be hardened in them, and they interfere less than the acid with the subsequent staining by dyes. But they do not preserve the minute structure of the tissues, at least of their cells and nuclei, as do solutions of chromic acid; nor is the hardening process so rapid (2 to 4 weeks). **Müller's fluid** is a $2\frac{1}{2}$ per cent. solution of bichromate of potash in water containing 1 per cent. of sulphate of soda.

Osmic acid.—This is the most rapid fixative for most tissues, and in some respects the most perfect, for it usually kills protoplasmic structures so suddenly that there is no time for any contraction or other change of form to occur. On this account it is extensively used, alone and combined with other reagents. But it has the disadvantage that, more than most other fixatives, it interferes with the subsequent staining of cells by dyes; nor does it bring out the fine structure of cells and nuclei as well as chromic or picric acid. It stains substances of a fatty nature black. It is usually allowed to act on a tissue only for a short time (15 minutes to 3 hours). It is conveniently kept as a 1 per cent. or 2 per cent. solution.

Acetic acid.—This is rarely used alone as a fixative, but has been a good deal employed in combination with other reagents. Acetic acid is also useful for determining the action of weak acids upon the tissues. For this purpose it may be kept in 1 per cent. solution.

Flemming's fluid is constituted as follows:—

        Osmic acid, 2 per cent. . . 4 parts
        Chromic acid, 1 per cent. . . 15 ,,
        Glacial acetic acid . . . 1 ,,

Hermann's fluid has a similar composition, but with 1 per cent. platinic chloride substituted for the chromic acid.

These solutions may be used undiluted, or they may be diluted up to ten or even twenty times with water. Tissues are hardened rapidly in them (2 to 5 days).

Picric acid.—One of the most valuable fixatives. It is used in saturated or half-saturated solution, either alone or with the addition of sulphuric or nitric acid to the solution in the proportion of 2 p.c. It hardens quickly, a day or two being usually sufficient. Its chief disadvantage is the difficulty of getting rid of the excess of acid from the tissue (this being inimical to the after-staining with dyes). With this object a tissue hardened in picric acid is placed for some days in 95 p.c. alcohol containing lithium carbonate in solution, the fluid being changed until the tissue yields no more colour to it. The tissue is then kept in alcohol. It should not be placed in water before alcohol.

Alcohol.—This is used in histology more than any other reagent. In strength of 70 per cent. upwards it serves both for fixing and hardening, the latter process being extremely speedy, especially when the alcohol is absolute. In a strength of 33 per cent. (one-third alcohol), as recommended by Ranvier, it is an invaluable dissociant. It has the great advantage that it rather assists than prevents the after-staining by dyes. It tends to produce a certain amount of shrinkage in some of the tissue elements, so that their intimate structure is as a rule less well displayed than after chromic or picric acid. But alcohol is almost invariably used to complete the hardening process when these acids have acted sufficiently long to fix the tissue elements, and it is also of universal employment for dehydrating preparations which have to be transferred from watery solutions to essential oils. The most useful strengths of alcohol to keep ready are 33 per cent., 50 per cent., 75 per cent., 96 per cent. (or methylated spirit), and absolute alcohol.

Formaldehyde has of late come into use as a fixing and hardening reagent. It is sold in a 40 per cent. solution, which is termed *formol*. It acts rapidly and penetrates freely, so that a small piece of tissue may be hardened in it in the course of a few minutes, and then transferred to alcohol. Formol is rarely used undiluted. For fixing and hardening purposes it is used in a strength of from 1 in 20 to 1 in 5, according to the nature and thickness of the tissue. It may be mixed either with water or with alcohol. Tissues as a rule stain well after formol, and its rapidity of action and penetrating power render it a most valuable reagent.

Corrosive sublimate is best used dissolved in 1 per cent. salt solution, as a saturated solution (which should be kept in the dark). It may be used either alone or with 1 g. picric acid added to each 100 c.c. of the sublimate solution. Mann further adds 2 g. tannic acid. Objects are fixed and hardened quickly, small objects in a few minutes, larger ones in a few hours. They are then transferred to 75 p.c. spirit deeply coloured by tincture of iodine, and after remaining in this long enough to remove the excess of sublimate they are transferred to stronger spirit, which should be changed once or twice. Objects hardened in corrosive sublimate usually stain readily with most dyes. If the blood-vessels are injected with it, after being washed clear of blood by normal salt solution, fixation and hardening is instantaneous. The tissues can then be placed in a quantity of the same fluid and transferred in a few hours to rectified spirit containing iodine. They are then passed through stronger alcohol to absolute.

FLUIDS USED FOR STAINING HISTOLOGICAL OBJECTS [1]

A piece of a tissue or organ may, after having been fixed and hardened, either be stained in bulk, or teased preparations or sections may first be made from it, and these may then be subjected to the action of a staining fluid. Bulk staining usually requires several hours' immersion in a weak solution of a dye, but sections may usually be stained in a few minutes

---

[1] Fluids and methods of staining for special purposes are given in the descriptions of the preparation of the several tissues and organs.

in stronger solutions. The following staining fluids are in most frequent use :—

Hæmalum (Mayer).—This is a solution of hæmateïn—the colouring principle of hæmatoxylin [1]—in alcohol, added to a solution of alum in water.

Ammonia alum	. 50 grammes	Hæmateïn	. . 1 gramme
Distilled water	. 1,000 c.c.	Rectified spirit	. 100 c.c.

Pour the hæmateïn solution gradually into the alum solution. Add a small piece of thymol to prevent the growth of moulds. This is the stock solution. It may be used in full strength when rapid staining or overstaining is desired, or it may be diluted to any degree with distilled water. It forms the most generally useful form of stain for sections, and is in constant requisition.

Delafield's hæmatoxylin.—Dissolve 4 g. hæmatoxylin in 25 c.c. absolute alcohol, and add the solution to 400 c.c. of a concentrated solution of ammonia alum in distilled water. Allow the mixture, which assumes a deep violet colour, to stand 3 or 4 days, and then add 100 c.c. of glycerine and 100 c.c. of methylic alcohol. The solution must be allowed to stand a few days before it will stain readily, but the results which are then obtained are extremely good. Both this solution and Mayer's hæmalum tend to lose their staining properties when long kept, and acquire a red colour, but their original colour and staining power may be restored by the cautious addition of ammonium sulphide. They should always be filtered before using, and diluted with distilled water only.

Acid hæmatoxylin (Kulschitsky).—Dissolve 1 gramme hæmatoxylin in a little alcohol, and add to it 100 c.c. of a 2 per cent. solution of acetic acid in water. Preparations after immersion in this fluid must be well washed with tap-water or lithium carbonate solution until the brown coloration is replaced by blue.

---

[1] In the following text the terms logwood solution, hæmatoxylin, or hæmateïn, are used indifferently for the colouring principle of logwood.

A similar stain is obtained by adding 2 parts glacial acetic acid to 100 parts hæmalum (Mayer).

Kleinenberg's hæmatoxylin.—This is useful for staining in bulk. It is prepared as follows :—

(A) Make a saturated solution of crystallised calcium chloride in 70 per cent. alcohol and add alum to saturation. (B) Make also a saturated solution of alum in 70 per cent. alcohol. Add (A) to (B) in the proportion of 1 : 8. To the mixture add a few drops of a saturated solution of hæmatoxylin in absolute alcohol.

It may, if required, be diluted with the mixture of A and B. The stained tissue can be placed at once in strong spirit.

Heidenhain's bulk stain.—This is one of the best stains for glandular organs. It is used as follows. Two solutions are prepared, A and B :—

A		B	
Hæmatoxylin	. 1 gramme	Yellow chromate	
Distilled water	. 300 c.c.	of potash	. 1 gramme
		Water	. 200 grammes

Dissolve the hæmatoxylin in a little alcohol and add the water. A piece of tissue after hardening in alcohol (or picric acid followed by alcohol) is placed in the hæmatoxylin solution (A) for 12 to 24 hours, and is then transferred for 12 to 24 hours to the chromate of potash solution (B).

Carminate of ammonia.—Dissolve one or more grammes of carmine in a small quantity of ammonia. Place in a porcelain capsule with a small piece of thymol and allow slowly to dry. Dissolve in water as required. Place the tissue or sections in acidulated water after staining.

Picrocarmine.—Dissolve 1 gramme of carminate of ammonia in 35 c.c. distilled water, and gradually add to it with constant agitation 15 c.c. of saturated solution of picric acid.

Mayer's carmalum.—For sections or bulk-staining.

Carminic acid (Grübler's)	1 gramme
Alum	10 grammes
Distilled water	200 c.c.

Boil together, allow to cool, and filter. Add thymol to keep.

STAINING FLUIDS 21

This is an invaluable staining fluid. If sections which have been stained with it are placed in alcoholic solution of picric acid, all the effects of picrocarmine staining are obtained.

Aniline dyes.—It may be stated, as a general rule, applicable to all the aniline dyes, that the staining power of a given strength of solution varies inversely with the amount of alcohol present. Absolute alcohol therefore rapidly extracts most of these dyes from preparations which have been stained with a watery or dilute alcohol solution of them. Acid fuchsin forms an exception to this rule. The aniline dyes most commonly employed are the basic colouring matters *methyl-green*,[1] *methyl-blue, fuchsin* or *magenta, gentian-violet, methylene-blue*,[2] *methyl-violet, safranin, thionin, toluidin-blue,* and *vesuvin,* and the acid colouring matters *eosin, acid fuchsin* (*rubin S*), and *orange G*.

They are best kept in saturated solutions in 96 per cent. alcohol or in water. For purposes of staining, either the saturated watery solution is used, or a dilute solution is made by adding some of the strong alcoholic or watery solution drop by drop to distilled water in a watch-glass. To obtain a very intense stain, the following method is used (Loeffler) :—Add 30 c.c. of the saturated alcoholic solution to 100 c.c. of a 0·01 per cent. solution of caustic potash in distilled water. Or :— Add 1 c.c. of a 1 per cent. solution of caustic soda to 100 c.c. aniline water (freshly prepared by shaking up 5 c.c. aniline with 100 c.c. distilled water, and passing through a filter wetted with water), and in this either dissolve the dry staining material with repeated agitation or simply add to it some of the alcoholic solution of the dye.

[1] Methyl-green is essentially a stain for nuclear chromatin, and it differs from most other aniline dyes in the fact that it must always be used in acid solution (1 p.c. acetic acid).

[2] Methylene-blue is chiefly used in normal histology for the staining of nerve-terminations *intra vitam,* or at least perfectly fresh. The method of employing it for this purpose will be given later. It can also be used in stronger solution to show epithelial outlines, cell-spaces of connective tissue, &c., such as are exhibited by nitrate of silver (p. 103). It is fixed in the tissue by subsequent treatment with nitro-molybdate of ammonia. Methyl-blue is quite a different substance.

For karyokinetic figures and centrosomes of cells it is best, as recommended by Henneguy, to place the sections before staining in 1 p.c. solution of permanganate of potash for 5 minutes (after which they are to be thoroughly washed with water).

In using most aniline dyes it is customary to overstain and afterwards partially decolourise. Alcohol may be used for this purpose, with or without the addition of 0·1 to 2 per cent. hydrochloric acid. When acid is used it must always be completely removed by washing first with water or 80 p.c. alcohol containing a little lithium carbonate and then ordinary water or alcohol. The process of decolourisation is arrested by transferring the preparations to oil of cloves, oil of bergamot, or xylol. Acid alcohol is used also to decolourise sections overstained by hæmatoxylin. Alcohol rendered alkaline with caustic potash is used to decolourise tissues overstained by acid fuchsin.

Double and triple staining.—*Picrocarmine, magenta, hæmalum-eosin, methyl-blue eosin, toluidin-blue eosin, Ehrlich-Biondi stain.*—Two or three stains giving different results may be used for a preparation, and they may be applied successively or simultaneously. *Picrocarmine* is in fact such a double stain, since it dyes some tissues yellow and others red.

Magenta (fuchsin) also gives a double stain, some tissues being coloured purple and others red. For hardened tissues, which are to be prepared as sections after staining in bulk, a saturated solution in rectified spirit may be used (see Ossification, p. 131). For fresh tissues the following solution is recommended. Take ·05 gr. magenta and dissolve it in 10 c.c. of alcohol. To the solution add gradually 20 c.c. glycerine and then water to 100 c.c. The fresh tissue is to be mounted in this fluid, and the cover-glass soon cemented, to prevent evaporation of water and concentration of the glycerine, which would result in the removal of the stain from the tissue elements. Elastic fibres are stained intensely by this means.

Eosin-hæmateïn.—One of the simplest methods of double staining of sections is to place them first for 20 minutes in

1 per cent. water solution of *eosin*, and, after washing with distilled water, in *hæmalum* until stained. Another way is to stain with hæmateïn first, and afterwards with alcoholic solution of eosin. From this they are passed directly into clove oil. Orange G may be used in place of eosin.

Methyl-blue and eosin.—This may be made as follows. Dissolve methyl-blue and eosin to saturation in water, filter, and add a little thymol to the solution. Use the stain either as it is or diluted with water from 5 to 20 times, according to the nature of the tissue and the manner in which it has been fixed and hardened. The time of staining will also depend upon these factors: thus, with some sections half a minute will be enough; with others, differently hardened, a much longer time, even hours, may be necessary. The sections may be overstained and decolourised. This last-named process is effected by absolute alcohol, or, more rapidly, by alcohol containing 1 or 2 per cent. HCl. The sections are then passed through pure alcohol into an alcoholic solution of eosin (1 per cent.), where they remain for a few minutes; they are then rapidly dehydrated in absolute alcohol and transferred to oil of cloves, oil of bergamot, or xylol.

Toluidin blue and eosin stain (Mann).—Place the preparation for ten minutes in a 1 per cent. water solution of eosin; rinse with water and place for twenty minutes in a 1 per cent. solution of toluidin blue. Decolourise just sufficiently and dehydrate in absolute alcohol, transfer to xylol and mount in xylol balsam. This is a very good general method.

Ehrlich-Biondi triple stain.—This, as modified by Heidenhain, is made as follows. Saturated aqueous solutions of orange G, acid fuchsin, and methyl-green are prepared, filtered, and mixed as follows:—To 100 c.c. of the orange G solution 20 c.c. of the acid fuchsin is gradually added, and then to the mixture 50 c.c. of the methyl-green. For actual staining this solution may be diluted with water up to 100 times its volume. Before being used it should be faintly acidulated with acetic acid, sufficiently to produce a well-marked carmine tint in a very dilute solution. Sections should be left for 24 hours in the stain. Preparations may be overstained by it, and decolourised by alcohol. It succeeds best with sections from sublimate preparations.

Microchemical stain for iron (Macallum). The sections (of alcohol-hardened tissues) are placed for a short time in an aqueous solution of pure hæmatoxylin (1 to 300). Any free iron in the tissue is at once stained black by the reagent. To show the presence of iron which is in organic combination the sections require to be previously treated for a short time with 96 p.c. alcohol containing 10 parts hydrochloric acid per cent.

Microchemical stain for phosphorus (Lilienfeld and Monti). The sections (fresh or hardened in alcohol) go first into a solution of nitro-molybdate of ammonia (a few minutes to several hours); they are then thoroughly washed and placed for some minutes in pyrogallol solution. They are then again washed and transferred through absolute alcohol and xylol into balsam. Parts containing phosphorus are stained brown.

FLUIDS USED FOR CLEARING AND MOUNTING SPECIMENS OF TISSUES AND ORGANS, AND FOR FIXING THE COVER-GLASS OF 'WET' SPECIMENS

Glycerine, either pure or diluted with an equal amount of water or in the form of **glycerine jelly,** is a most valuable medium for mounting permanent histological specimens. Its high, but not too high, refractive index enables it to render clear many of the details of an object which might from the otherwise too great opacity of the tissue remain obscure. At the same time it is desirable for some objects that this clearing should be tempered; a result easily attained by diluting the glycerine. Dilution is also necessary in the case of specimens stained with aniline dyes, most of which would be dissolved out by strong glycerine. In these instances, where diluted glycerine is employed for mounting, it becomes necessary to prevent evaporation of the water, and with this object the edges of the cover-glass are surrounded or 'ringed' with some fatty or resinous material For temporary purposes a film of *olive oil* or molten *paraffin* led round the edge of the cover-glass is sufficient, but for permanent preparations the best fixing fluid for the cover-glass is **gold size**, which is painted round with a small camel-hair pencil. For the fixation to be successful it is necessary that the surfaces of the cover-glass and slide to which the gold size is applied should be dry and clean, and not wetted by the glycerine used in mounting. When glycerine jelly (which must be warmed for use) is employed the cover-glass need not be 'ringed.'

Canada balsam dissolved in xylol (**xylol balsam**) is the other chief fluid used for mounting, and at the same time 'clearing,' preparations. The refractive index of this substance is so high that unstained preparations and parts of stained preparations which have remained relatively little stained by the dyes used are rendered perfectly clear and transparent, and the details of structure in these are almost invisible, thereby showing up by contrast the details of the stained portions of the tissue. Xylol balsam is chiefly used for the mounting of stained sections. As it is absolutely immiscible with water all specimens which are mounted in it must first be completely dehydrated. This is effected by immersing them in absolute alcohol. Alcohol, however, is also immiscible with balsam, and the specimens must therefore first be passed through an essential oil which will mix with either alcohol or balsam. The one most commonly used is **oil of cloves**, and this has the advantage that the alcohol used for dehydration need not be absolute, for the oil of cloves will take up a little water. Other essential oils which are employed are **oil of bergamot, oil of cedar, oil of turpentine, and xylol.**

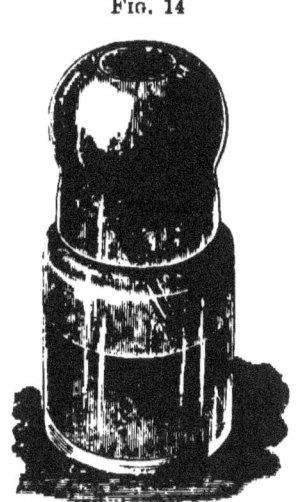

Fig. 14

Bottle for xylol balsam

Dammar varnish is sometimes used in place of xylol balsam. It is made by dissolving dammar resin in a mixture of equal parts of warm benzine and turpentine; the solution is to be filtered through paper wetted with chloroform.

The xylol balsam or dammar varnish is best kept in a bottle such as that shown in fig. 14. It is provided with a cap, ground to fit accurately to the neck of the bottle, and with a loose glass rod for dipping out the solution.

METHODS OF PREPARING SECTIONS

After the preliminary processes of fixing and hardening are completed, sections can be prepared from a tissue or organ. It

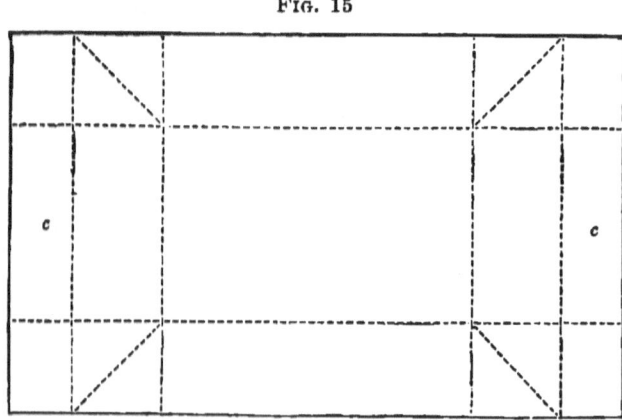

FIG. 15

Outline showing the manner in which a small piece of paper is to be folded to make an embedding trough

is now usual to cut them with the aid of some sort of microtome, but for certain purposes it is still desirable to cut by hand with a razor, which is usually then wetted with spirit. The piece of tissue to be cut is, if large enough, held in the left hand; if small it may be grasped in a split cork or in a piece of alcohol-hardened liver, or it may be fixed in a small trough of paper or lead foil (figs. 15, 16, 17) and molten paraffin

PREPARATION OF SECTIONS 27

poured in so as to enclose it completely. The razor must be held horizontally and flooded with spirit (fig. 18).

For cutting sections with a microtome it is desirable to securely fix the tissue by allowing it to be permeated with some material which will set to a consistence suitable for slicing.

FIG. 16

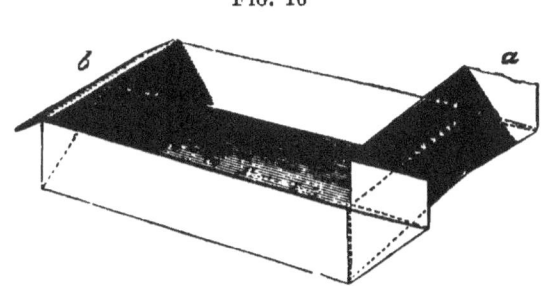

Embedding trough made from a piece of paper of the size shown in fig. 15, with one side, b, completed; the other, a, only half finished, so as to show the manner in which the corners are folded and fixed

FIG. 17

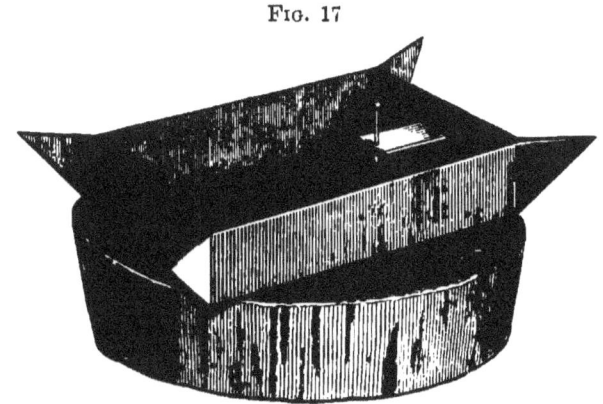

Embedding trough of lead foil placed on a cork, and with a piece of tissue *in situ*. All that is further necessary is to fill the trough with molten paraffin

The materials which are commonly used for this purpose are gum (frozen), collodion or celloidin, and paraffin, and the methods are termed respectively the freezing, the celloidin or collodion, and the paraffin methods.

28 PRACTICAL HISTOLOGY

Freezing method.—This is by far the most expeditious method of preparing sections, and for many histological purposes is as good as any other.

A thin piece of the hardened tissue is first soaked in water, then placed in a thin syrup of gum arabic or dextrine for a few hours, and then on the plate of a freezing microtome (fig. 19). On the under surface of this plate an ether spray is now allowed to play until the tissue is just frozen through ; it is

FIG. 18

Process of cutting sections of an embedded tissue

better that it should not be too hard frozen. Sections are now made with the plane-iron or razor provided with the microtome, the freezing plate on which the tissue lies being gradually screwed up as the successive sections are taken. The sections are placed in water to dissolve away the gum, and the best are then to be selected and treated with such staining and other reagents as may be desired. They can be readily mounted at once in glycerine, or they may be passed, after staining, through water, rectified spirit, alcohol, and oil of cloves, and

PREPARATION OF SECTIONS

mounted in balsam in the manner described below under the paraffin method.

Collodion method.—This is useful for preparing sections of large objects in which it is desirable to hold the several parts securely together during the staining and subsequent treatment of the sections. The organ must, after hardening,

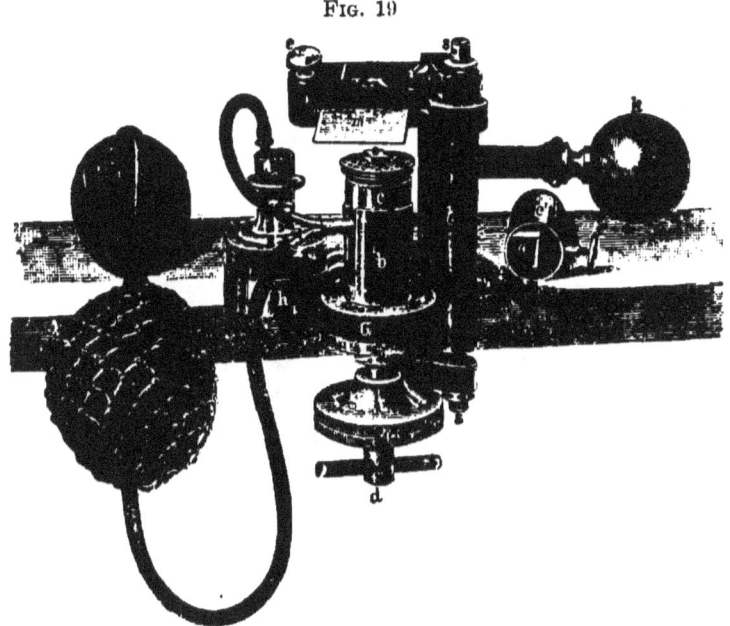

Fig. 19

Freezing microtome

o, object, soaked in gum and placed on metal plate of object-holder, *c*; *b*, box in which the ether collects after spraying on the under surface of the plate; *c'*, extra holder for paraffin-embedded tissues; M, micrometer screw for raising object; *m*, razor fixed in a clamp, and moving horizontally on the axis , *s s*, along with the pivotted bar *a*; *e*, screw for tightening clamp of razor; *k*, handle; G, stand of microtome fixed to table by clamping screw, *d*; *h*, holder for ether-bottle

be cut into moderately thin slices, and these are placed in absolute alcohol for a short time, and then in a mixture of equal parts of alcohol and ether. From this they are transferred to collodion made by dissolving 1 part of pyroxylin, or phytoxylin, or celloidin in 15 of the above mixture, *i.e.* more than double the strength of the collodion of the Pharmacopœia.

In this they are left in a well-stoppered bottle for some hours or days, according to their thickness. Each is then placed upon a brass holder of appropriate size and shape (fig. 20), and the collodion is allowed to set by evaporation of the ether. This will occur within a few minutes, and the whole is then immersed either in chloroform or in 85 per cent. alcohol. In this the collodion becomes, in a few hours, of a consistence suitable for the preparation of sections, which must be made with a microtome the knife of which is horizontal and is kept flooded with alcohol. It is sometimes preferred to soak the collodion block in cedar-wood oil (after dehydrating with 96 p.c.

Fig. 20

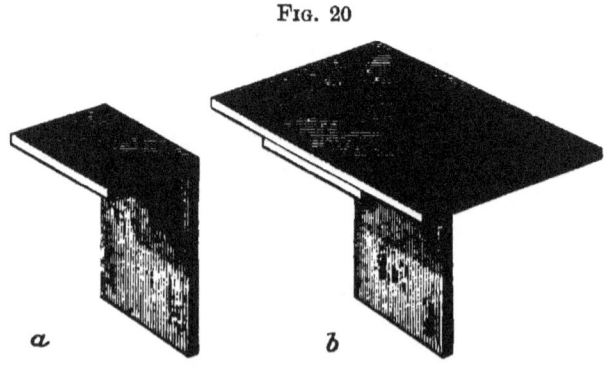

Brass holders for collodion-soaked preparations

a, for small pieces, is simply a piece of angle brass sawn off a strip ; *b*, for large pieces, is the same with a brass plate soldered to it

alcohol) and to cut with the knife wetted with the same fluid. The most suitable microtome for collodion-embedded tissues is the *inclined plane microtome* (fig. 21) : the knife must be set as obliquely as possible. If the tissue had not been stained in bulk the sections may now be stained, the methods employed being the same as for sections cut by the freezing microtome.

It is generally desirable to keep the collodion in these sections, so that they may be better held together. They must therefore not be placed in absolute alcohol, nor in oil of cloves, for the former softens and the latter dissolves collodion, but, after dehydrating in 96 p.c. alcohol, oil of cedar-wood may be

PREPARATION OF SECTIONS

used for clearing them, and they can then be mounted in xylol balsam.

Paraffin method.—This is the most generally useful, and for most purposes the best method of preparing thin sections of a tissue. Moreover, it enables any number of successive sections to be taken and kept in their serial order ; a point of fundamental importance in practical embryology and morphology, and of some importance in certain branches of histology.

The tissues to be embedded must always be ultimately placed, whatever fixing and hardening fluids have been used, in

Fig. 21

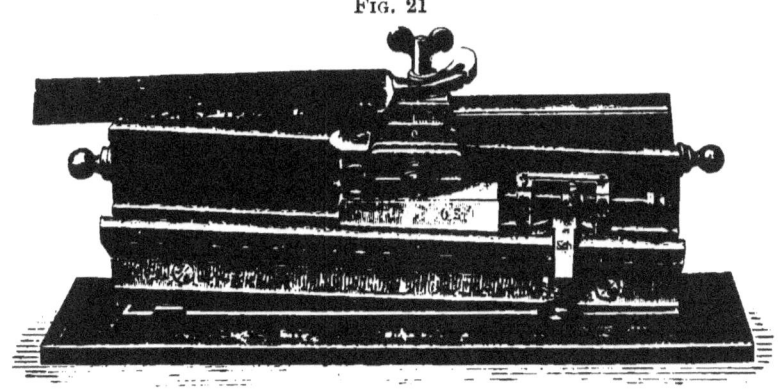

Inclined plane microtome

The object, o, is fixed in the clamp H, which is attached to the object slider, o s, sliding up an inclined plane, the exact amount of movement being determined by the micrometer screw, J. The knife is fixed upon another slider which moves horizontally at the back of the instrument

strong alcohol. It is advantageous that they should be stained in bulk, as this saves the somewhat tedious process of staining the individual sections, but it is not always possible to effect bulk-staining satisfactorily. From alcohol the tissue, in sufficiently thin pieces, is placed in oil of cedar-wood until soaked through.[1] To facilitate the penetration the oil of

[1] Chloroform, turpentine, and xylol may be used instead of oil of cedarwood; they must be preceded by absolute alcohol. For delicate objects the transference from alcohol to paraffin is made gradually through *alcohol and chloroform*, *pure chloroform*, and *paraffin* dissolved in *chloroform*, the chloroform being finally driven off by gentle heat; but this long process is rarely necessary in vertebrate histology.

cedar-wood may be gently warmed (to 40° C.). After it has completely penetrated, the piece, if small, is transferred to a shallow glass vessel (*e.g.* a watch-glass) which has been smeared with glycerine, and into which paraffin of melting point 45° to 50° C. has been poured, and kept just molten in a suitably regulated oven or water-bath. In this it is left for half an hour or several hours, according to size. The vessel is then removed and cooled as quickly as possible. When hard set the whole is turned out of the watch-glass, and the excess of paraffin cut away, leaving the tissue in the middle of a small square block of paraffin. If the piece is larger, it is placed for some hours in a larger vessel of melted paraffin,

Fig. 22

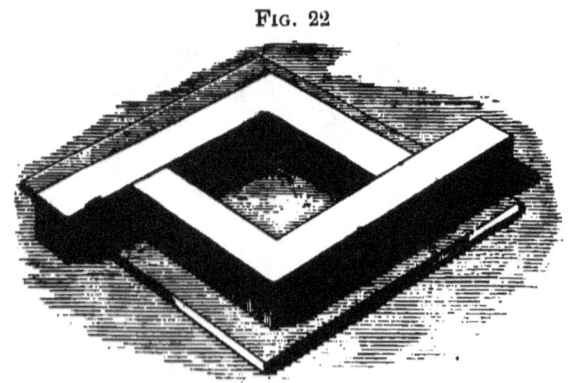

Mould for paraffin

and this is then poured into a mould made of paper or of lead foil (figs. 16, 17), or by placing together two L-shaped pieces of brass (fig. 22), the tissue being arranged with warmed needles within the paraffin, or fixed in place beforehand by a pin (fig. 17). In either case the result is that the tissue which has been soaked with paraffin is also completely surrounded and supported by paraffin of the same consistence, and is thus in a position to offer a uniform resistance to the cutting instrument.

The block of paraffin is now fixed in the desired position in the holder of a microtome. Either of those which have been already mentioned for freezing or for celloidin will serve for cutting paraffined tissues, but the knife must always be used dry, and set exactly transversely. The three instruments immediately to be described are, however, better adapted for the purpose.

The tripod microtome (fig. 24).—This consists of a small frame of brass or cast iron to the under surface or one end of which an ordinary razor can be clamped. Of the three feet of the tripod, two are fixed and one is provided with an adjustable screw. The microtome frame is moved by the hand over a flat piece of glass upon which the paraffin block containing the tissue has been attached by aid of heat, and successive sections

FIG. 23

Box for storing specimens enclosed in paraffin blocks

are cut by lowering the cutting edge of the razor by a slight turn of the screw adjustment, and sliding the holder over the glass. This is not only a very efficient, but also a very cheap form of microtome, for the frame can be made for two or three shillings.[1]

If it is not required to keep the sections in series they may be placed, as soon as they are cut, in turpentine or xylol, which quickly dissolves the paraffin. If the tissue has been stained in bulk all that is now necessary is to mount one or more sections in xylol balsam. But if the sections require to be stained they are passed first from xylol into absolute alcohol,

[1] Made by A. Kershaw, Cankerwell Lane, Leeds.

then into rectified spirit, then into distilled water, then into the stain, *e.g.* solution of hæmalum. After staining they are transferred successively to water, rectified spirit, absolute alcohol, and xylol, and are at length ready to be mounted in xylol balsam. These transferences are best effected in glass or earthenware pots, provided with well-fitting covers (fig. 25), and the sections should be left at least three minutes in each fluid.

Fig. 24

Tripod microtome (Birch's pattern)

The preparation of serial sections. The rocking microtome.—In this instrument (fig. 26) an ordinary razor is also used, but it is fixed vertically with the edge uppermost, and the tissue is moved over it, the paraffin block containing the tissue being attached to a lever which is worked up and down, and also gradually thrust forward, as the handle, H, is worked. In this way successive sections are cut automatically of uniform thickness. If the sides of the paraffin block are parallel, and the paraffin not too hard, the successive sections adhere to one another edge to

Fig. 25

edge into a ribbon, which may hold together for a considerable length. Such paraffin ribbons may be cut by other microtomes, using the knife dry, and with the edge at right angles to the

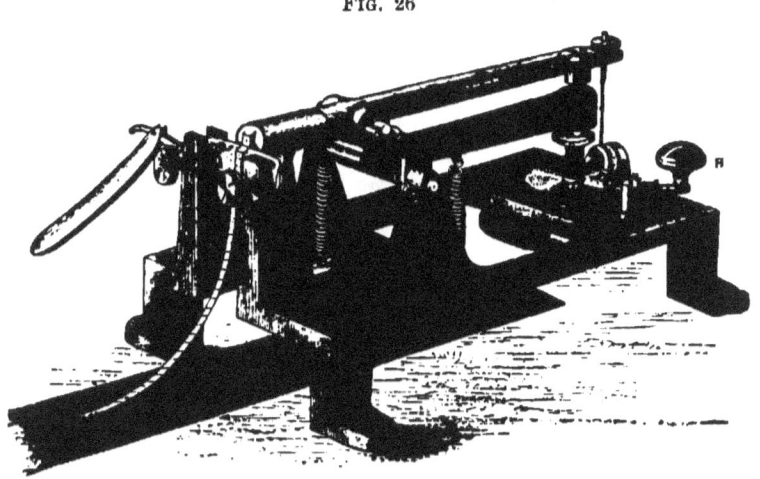

Fig. 26

Rocking microtome (Cambridge pattern)

direction of movement, but instruments like the rocking microtome, which have the razor fixed vertically, are the most convenient forms for this purpose.

D 2

Minot's microtome.—In this instrument the razor is also fixed vertically, and the block of paraffin containing the embedded tissue is caused to move vertically up and down over the edge of the razor, with an alternate movement of advance, which is capable of being regulated by a screw adjustment to the utmost degree of nicety.[1]

Methods of fixing paraffin-cut sections in series upon a slide.—The following methods are used for fixing the series

Fig. 27.

Minot's microtome

or ribbon of paraffin-cut sections upon a slide, and thus keeping the sections in order whilst being treated for permanent mounting:—A ribbon of sections having been cut of a length suitable for mounting, it is arranged appropriately upon a

---

[1] Both the Cambridge Scientific Instrument Co. and Messrs. Bausch & Lomb (for Prof. Minot) have recently brought out new microtomes, as improvements upon those here described; but they are larger and much more expensive, and the above will be found to answer every practical purpose.

carefully cleaned slide which is freely wetted with water. The slide is then gently warmed until the sections are flattened out, when the excess of water is drained off, and the slide allowed to stand in a warm place until the water has *completely* evaporated, leaving the paraffin ribbon adhering to the glass. The slide is now heated until the paraffin just melts; it is then allowed to cool and is immersed in xylol. This dissolves the paraffin, but the sections are left sticking to the slide. If the tissue was stained in bulk, and the sections are therefore already sufficiently stained, they can be at once mounted in xylol balsam and covered.

Various solutions, such as collodion dissolved in oil of cloves and shellac dissolved in creosote, are employed with a view to fixing the sections more securely. They are chiefly necessary for chromic-hardened tissues. In any case it is well to combine them with the water method, for it is only by the latter that a series of paraffin sections can be properly flattened out. They should be brushed over the sections when they are completely dry, and before the paraffin has been dissolved by xylol. Diluted white of egg is also used very frequently for the same purpose. A solution which can be kept for some time is made by shaking up 50 c.c. egg-white with 50 c.c. glycerine and dissolving 1 gramme salicylate of soda in the mixture. The slide which is to receive the sections is smeared thinly over with this, then covered with water, and the series of sections is laid in position. It is now gently warmed until the paraffin is softened and the sections lie flat, when the excess of water is removed by draining and evaporation. The slide is then immersed for an hour in a mixture of alcohol and ether to coagulate the albumen, then in xylol to dissolve away the paraffin, and finally the sections, if already stained, may be covered in balsam.

But if not stained in bulk a further procedure is necessary. The slide, after immersion in xylol, is placed in absolute alcohol for a few minutes, then in 50 per cent. spirit, then in the staining fluid, *e.g.* hæmalum. After the sections are sufficiently stained the slide is transferred successively to water, 50 per cent. spirit, absolute alcohol or methylated spirit, and xylol or oil of bergamot or oil of cloves,[1] and finally the sections are covered with xylol balsam and mounted.

Special oblong porcelain vessels with grooved ends are made for holding the series of fluids required for effecting these transferences (fig. 28); they should be provided with well-fitting covers. The slides are placed bodily into them. The series of operations may be performed, if desired, by pouring the successive fluids freely over the sections. They may also be performed with the sections fixed beforehand to a cover-glass instead of upon a slide. The following diagram may serve to show at a glance the several transferences which are necessary in order to stain and permanently mount paraffin-cut sections :—

DIAGRAM TO SHOW METHOD OF TREATMENT OF PARAFFIN-CUT SECTIONS

1. Place on a slide in water : warm gently.
2. Drain off water : allow to dry completely.
3. Warm until paraffin is just melted.
4. Xylol.

If tissue is already stained in bulk.

Xylol-balsam

For sections cut by the freezing or by the celloidin methods, processes 1 to 6 of the accompanying table are omitted, if the tissue has not been stained in bulk. If, however, it has been already stained in bulk, the sections need only be put through the last three of the series.

If sections require to be stained.

5. Absolute alcohol.
6. 50 per cent. alcohol.
7. Stain (*e.g.* hæmalum).
8. Water.
9. 50 per cent. alcohol.
10. Absolute alcohol or methylated spirit.
11. Xylol or bergamot oil or clove oil.
Xylol-balsam.

---

[1] Xylol needs to be preceded always by absolute alcohol, but as this is somewhat expensive methylated spirit may be used instead, in which case oil of cloves or oil of bergamot must be used as the intermediary between alcohol and Canada balsam.

GENERAL DIRECTIONS FOR WORK. METHODS OF DRAWING, PHOTOGRAPHING, AND MEASURING MICROSCOPIC OBJECTS

Before commencing see that the table is in order and clean, and that everything is at hand that is usually wanted for histological work. Especially look carefully to the glasses of the microscope that there is no dust or other impurity on them. If any glycerine or Canada balsam should have found

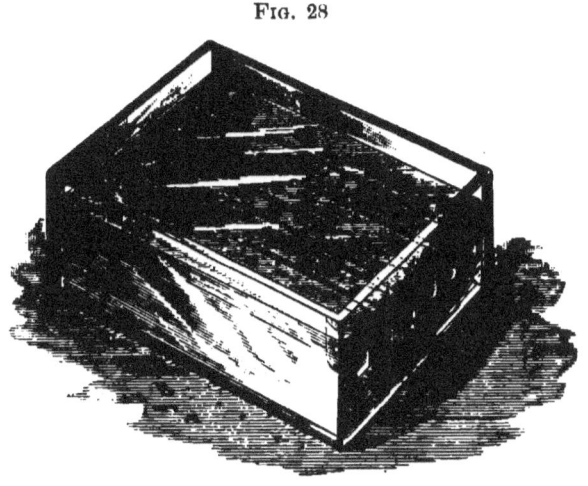

FIG. 28

its way on to the objective, as is often the case when sufficient care is not taken in placing a preparation upon or removing it from the stage, they are to be rubbed off, the former by a cloth wetted with water, the latter with a little spirit. Before beginning to prepare a tissue it will be necessary to look over the description of the mode of making each preparation in order to know what vessels and what reagents to get together. Otherwise many a specimen will be spoiled by being left too long in one fluid, whilst the one to which it should be transferred is being got ready. The cover-glass should always be cleaned and dried before commencing, and placed ready to hand in some situation where it is not likely to get broken. It is well always to put the cleaned cover-glasses in the same place—say on the foot of the microscope or in special clips (fig. 8)—otherwise when wanted quickly it is often difficult to find them.

Every specimen that is to be kept must be distinctively labelled as soon as made ; and if there is anything of importance to remember about it this must be at once entered in the notebook, which no one who is working with the microscope should be without.

Modes of drawing microscopic objects.— The student should never trust to the transient impression of form or

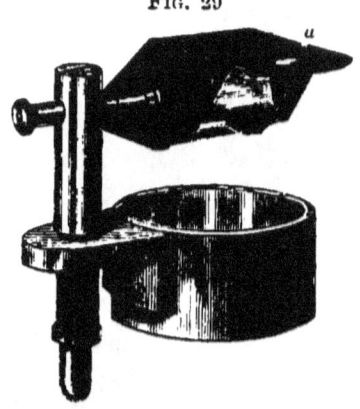

Fig. 29

Camera lucida, for tracing the outlines of an object without tilting or otherwise disturbing the microscope. The metal ring fits on the upper end of the microscope tube, and the aperture, a, is placed immediately over the eyeglass, this part of the camera being somewhat more depressed than is represented in the figure

structure which the mere glance at a microscopic preparation conveys, but should always, whether naturally a good draughtsman or not, endeavour to perpetuate the impression so obtained by a careful sketch showing the more important points which the preparation illustrates. Even without skill a little practice soon enables a sketch to be produced which gives a fairly good idea of the appearances seen, and, however rough it may be, serves materially to assist the memory.

But if greater accuracy of delineation is desired the outlines may be traced with a camera lucida. The simplest for ordinary use is that of Abbe, but the principle of all is the same, viz. to throw together upon the retina an image of the object under examination and of the paper on which the drawing is to be made. The image of the object then appears projected upon the paper, and its outlines may be followed with the pencil. In Abbe and Zeiss's instruments the surface of the paper is reflected by mirrors to the eye which is applied to the microscope, the object being viewed directly; in some other forms the paper is viewed directly and the image of the object by reflection. In using the camera neither the paper

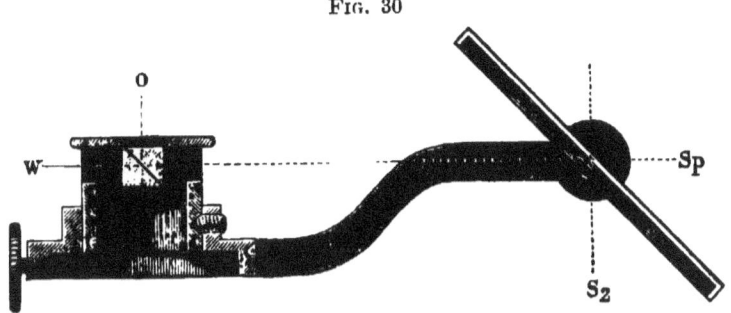

Fig. 30

Camera lucida of Abbe

The eye looking down the microscope is placed at o, and the image of the paper is reflected by the mirror and prisms along the lines s_2, sp, w, o, to the eye

nor the object should be too brightly illuminated, and the paper should not be placed on the table, but on a box or block at about the level of the stage of the microscope. It is not possible to trace the finer details with the camera, but only the broader outlines of objects.

Modelling.—The modelling of microscopic objects is required more in morphological than in histological studies, but a word here as to the method may, nevertheless, not be inappropriate. The object is to obtain a model of an organ reconstructed from serial sections. The object is attained by tracing upon wax plates of uniform thickness the outlines of the organ as they appear in sections taken at successive levels; the wax plates then have the superfluous parts cut away, and when piled upon one another in series a magnified model of the organ is obtained.

Fig. 81

Micro-photographic apparatus. One-seventh the natural size

MICRO-PHOTOGRAPHY 43

Fig. 32

Horizontal micro-photographic apparatus

The camera-box is fitted with a side door to enable the operator to observe the image upon a white screen on the camera-back whilst manipulating either the preparation or the focus of the microscope. The focussing can also be effected from the end by aid of the handle and rod shown in the figure. The board upon which the microscope and the source of light are placed can be rotated so as to bring the microscope away from the camera and enable a direct observation of the object to be made. The bellows, which are supported in the centre to prevent sagging, give a range of from 1 foot to 3¼ feet.

In order to maintain the proper relative position of the outlines one above the other the paraffin block must be marked with lines at right angles to the planes of section, and this can be done by cutting clefts longitudinally in it, along one or more sides, and filling them with lampblack or vermilion. The sections of these marks are also sketched in upon the wax plates, and indicate exactly how the plates are to be superposed. The plates can be stuck together by warming their edges.[1]

Micro-photography.—Certain microscopic objects lend themselves better than others to photographic reproduction, but it is in any case only the thinnest and best specimens which yield results at all commensurate with the time and pains it is usually necessary to expend to obtain really good results in micro-photography. The method employed needs no special description. An ordinary camera may be used, the lens being removed, and in place of it the upper end of the microscope tube—which must be placed horizontally for this purpose—inserted, movably but light-tight. But it is better to have a special camera made with a body at least twice as long as usual. It may be arranged either vertically or horizontally (figs. 31 and 32). Apochromatic objectives should be employed if possible, and they may be used either without eye-piece or with a projection eye-piece, whereby it is easier to obtain a flat field, although with considerable loss of light. The best source of light is either an arc or an oxyhydrogen light, the substage condenser being used for high powers. For low powers the light may advantageously be used without a condenser. In all cases, and certainly when a condenser is employed, it is necessary to employ an alum- or a water-cell to cut off the heat rays. The best dry plates for micro-photography are the isochromatic, preferably with antihalation backing.

Micro-photographs can rarely take the place of drawings, never certainly for the student. But they are often useful for assisting the draughtsman, and for this purpose no form of print is so convenient as those upon ferro-prussiate paper. These are printed in sunlight and developed, and fixed simply by washing with water. When dry, the print, including just so much detail as is desired,

---

[1] Full details of the method of modelling from sections will be found in G. Born, 'Die Plattenmodellirmethode,' *Arch. f. mikr. Anat.* xxii. p. 584; H. Kastschenko, 'Methode zur genauen Reconstr. u.s.w.,' *Arch. f. Anat.* 1886, p. 388; and H. Strasser, 'Ueber die Methoden der plast. Reconstruction,' *Zeitschr. f. wiss. Mikr.* iv. p. 168. Also in a later paper by Born in *Zeitschr. f. wiss. Mikr.* v. p. 433.

may be traced over with a lead pencil, and the blue colour then removed by immersion in carbonate of soda solution, followed first by dilute (1 per cent.) hydrochloric acid and then by water. This method of delineating microscopic objects will, in many cases, be found extremely practical.

Methods of measuring objects under the microscope.—If while the one eye looks down the tube the other is allowed to remain open, an image of the object will appear projected on the table at the side of the microscope, and it is not difficult to mark off, upon a sheet

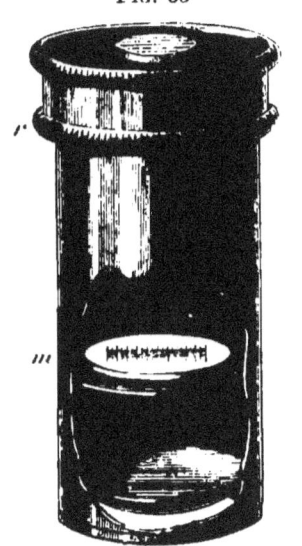

Fig. 33

Ocular micrometer, natural size

Part of the side is represented as broken away to show the field-glass at the bottom, and the micrometer-glass, m, a little below the middle. The collar, r, serves to vary the distance of the eye-glass from the micrometer

of paper placed there, the points between which the measurements are to be taken. The preparation is then removed, and a stage micrometer is substituted for it, the parts of the microscope being left in the same condition as before. The *stage micrometer* is a glass slide on which fine equidistant parallel lines have been ruled with a diamond. They are usually ruled transversely to the long axis of the slide, but it is better to have them parallel to the long axis. The distance between the lines is marked on the slide; it is generally either the $\frac{1}{100}$th and $\frac{1}{1000}$th part of an inch, or the $\frac{1}{10}$th and $\frac{1}{100}$th part of a millimetre. The lines are observed with the micro-

46　　　　　　　PRACTICAL HISTOLOGY

scope in the same way as the object, and their image can of course be similarly projected upon a sheet of paper and there marked down. The distances between the lines being known, it is easy, by comparison of the two markings, to find out the distance between the opposite points of the object. The projection of the lines of the stage micrometer can be still more easily effected with the aid of a camera lucida.

FIG. 34

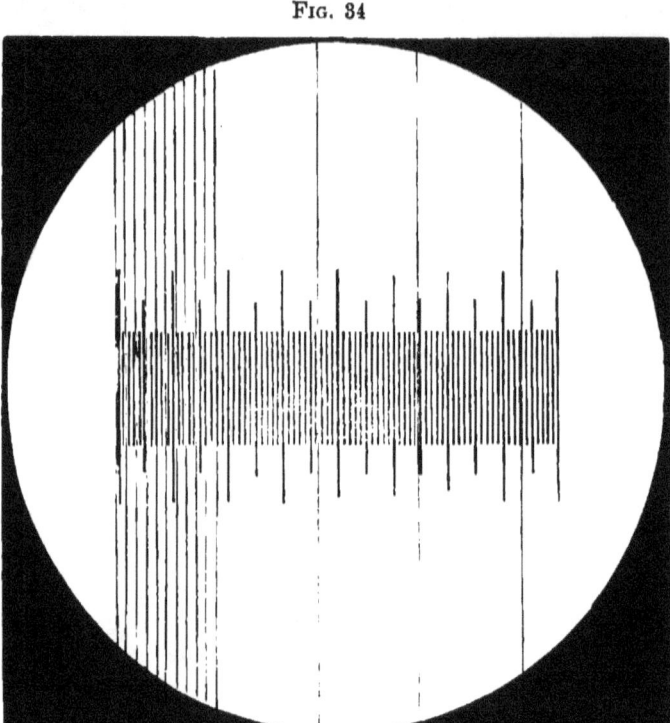

Lines of stage micrometer viewed with an ocular micrometer

The finer lines are those of the stage micrometer; about eighteen of the lines of the ocular micrometer are comprised in one of the larger intervals between them, so that, if these intervals represent $\frac{1}{100}$ inch, the subdivisions of the ocular micrometer will represent $\frac{1}{1800}$ inch

The microscope may be provided with an *eye-piece micrometer* (fig. 33). This is an ordinary ocular with a flat piece of glass (*m*), having a scale ruled upon it by a diamond, inserted between the field-glass and eye-glass. The value of the divisions of the scale should be determined once for all for each objective by observation of a stage micrometer (see fig. 34), the tube of the microscope being fully drawn out, and should be marked on the ocular; in subsequently using it for measurement all that is necessary is to see how many divisions of the scale the object under examination covers. Thus, supposing it had been found by examination of a stage micrometer that with the high-power objective and the tube drawn out each division of the eye-piece micrometer was worth $\frac{1}{1800}$ inch, any object which when viewed by the same objective and length of tube took up three divisions of the eye-piece micrometer would measure $\frac{3}{1800}$ths or $\frac{1}{600}$th of an inch.

The advantage of the eye-piece micrometer is that when its values are once ascertained the size of an object can be read off at once.

Determination of the magnifying power of a microscope.—The magnifying power of a microscope is determined by comparing the distance between the lines of the stage micrometer, as they appear imaged upon the paper, when this is exactly *ten inches* [1] from the eye, with the known interval between them. For instance, if, with the high-power objective and the ordinary ocular, the interval of $\frac{1}{1000}$th of an inch of the micrometer was represented on the paper by a space of half an inch, this interval is magnified as many times as the $\frac{1}{1000}$th of an inch will go into half an inch, that is to say 500 times; and every other object under similar conditions is magnified to a like extent.

The enlargement thus obtained may be determined once for all for each objective, the same ocular being used and the tube being drawn out to the same extent—say to the full length—in each case, and scales may be made representing the intervals between the micrometer lines under the different powers. For purposes of measurement it will then only be necessary to compare the projected image of an object with the scale which was made under like conditions, without again making use of the stage micrometer.

[1] This distance is taken arbitrarily.

CHAPTER I

THE BLOOD

Examination of human blood.—A drop of blood may be most conveniently obtained for examination from the finger. It is generally sufficient to give the end of the forefinger of the left hand a smart prick with a clean needle, in the thin part of the skin adjoining the root of the nail, squeezing firmly with the right hand above the point pricked to cause a drop to exude. If necessary, the finger may first be congested by tying a piece of string tightly round it. As soon as a small drop of blood has been pressed out, take up a previously cleaned cover-glass by one edge with forceps, let the drop come in contact with the lower surface of the glass near the opposite edge, so that a little adheres ; and then, letting this edge come first in contact with the upper surface of the slide near its middle, gradually lower the other edge, which is still held in the forceps, on to the slide (fig. 6, p. 8). When the lower blade of the forceps nearly touches the slide, withdraw the instrument carefully, so that the cover-glass may now rest evenly upon the slide by its whole under surface, with the blood in a uniform thin layer between. It is important not to let the cover-glass down too suddenly, for if dropped carelessly on the slide many of the corpuscles will be broken and destroyed. When the cover-glass is in its place, there ought to be just enough blood *entirely* to fill the space between the two glasses ; but it is better to have too little than too much. It might be supposed that the delicate corpuscles would be crushed between the cover-glass and slide, but they are for the most part protected from this by the buoying-up of the cover-glass by the liquid in which they float. With a little practice the cover-glass can be placed in position equally well without the use of forceps by holding its opposite edges between the finger and thumb.

The preparation made, it is to be at once transferred to the stage of the microscope, and examined with a power of from 300 to 500 diameters. The field of the microscope will be seen crowded with corpuscles floating in a clear liquid. Probably the first thing which will strike the beginner is the very faint colour which the so-called *red* corpuscles present, and probably these will also be the only kind of corpuscle that he will at first be able to distinguish. But if at the moment of observation there happens to be a current in the fluid—produced either accidentally by a shaking of the room or a draught of air, or purposely by gently touching the cover-glass with a bristle—it will be seen that while most of the corpuscles are carried along by the current two or three remain sticking to the glass, whilst the others are carried past them ; and on close examination it will further be clear that these are of a different nature from the rest, being entirely devoid of colour and of a pale, granular appearance. They are, in fact, white corpuscles, and once seen will be easily recognised again, even when the fluid is at a standstill.

Another thing that will be made manifest by any motion in the fluid is the biconcave discoid shape of the red corpuscles, for as they roll over it will be seen that their outline is no longer circular as when lying flat, but that when a lateral view of the discs is obtained the flattening and incurvation of the surfaces become evident.

When the motion in the layer of blood, in whatever way it may have been produced, is subsiding, it will be seen that whenever one corpuscle comes in contact with another the two seem to be in some way attracted to one another, so as to adhere closely by their opposed surfaces ; and other corpuscles

E

coming in the same way in contact with these and adhering, little piles, or rouleaux, are thus produced, which form by their junction with one another a network, extending throughout almost the whole of the preparation. In the cords of this network nearly all the red corpuscles are involved, and now for the most part are seen edgeways; but in other parts of the preparation where the layer of blood is very thin —the space being too small to allow the corpuscles to stand edge up, and to combine so completely to form rouleaux—they may be found still lying flat and distinct from one another; and these more isolated corpuscles may now be subjected to careful examination. Keeping a single red corpuscle in view, if it be brought exactly into focus—that is to say, if the microscope be so adjusted that the contour of the corpuscle is as distinct as possible—it will be observed, with the power ($\frac{1}{6}$ inch) which is at present being employed, that the middle part appears slightly darker than the rim, whereas if, by means of the fine adjustment, the objective be now brought somewhat *nearer* (lower), the middle part will come to appear lighter.

The cause of this is probably to be found in the *shape* of the corpuscle, the middle part of which acts upon the light like a biconcave lens, refracting the rays of light which are transmitted through it slightly outwards; so that if the objective is at a certain distance, all of the rays which traverse the central part do not reach it, some of them being deflected too much to the side to impinge upon the lower glass of the objective. The part in question, therefore, looks a little dimmer than the somewhat convex marginal part, whereas when the objective is brought *nearer* all the rays are intercepted by it, and the middle part, owing to its greater thinness, appears lighter than the rim. If an objective of very short focal length is used, the middle part of the corpuscle will be the lighter, even when the focus is rightly adjusted, for such an objective approaches near enough to intercept the outwardly refracted rays.

With the exception of these differences of shading (which are merely dependent upon the shape of the corpuscle), the red corpuscles present a perfectly homogeneous appearance, and exhibit in the fresh condition no tendency to separate into the two parts of which, as the study of the action of reagents will show, they in reality consist. But there may generally be noticed, even in a preparation which has been made with the greatest care, a red corpuscle here and there which varies from the prevailing form, having become more globular, and at the same time rather smaller in diameter.

These retain a smooth contour, whilst others, especially those near the edge of the preparation, have a jagged or crenate margin, as if set with little projections, and such projections may also be seen on the surfaces of the corpuscles by carefully adjusting the microscope. This change of form, which is very characteristic of the mammalian red corpuscles on exposure, seems to be generally caused by a shrinking of the corpuscles, induced by an increase in the density of the plasma in which they float : it may always be produced by adding salt to blood.

Turning now our attention to the white corpuscles, not more than two or three of which are to be seen in each field of the microscope when the ordinary high power is being used, and which, as before stated, are readily distinguishable from the red corpuscles by their want of colour and their pale, granular aspect, we usually notice, if seen soon after the blood is drawn, and provided they are not pressed down by the cover-glass, that they are spheroidal and completely motionless, exhibiting no indications of vitality. Some of them may be noticed to contain a small group of well-marked granules, much coarser than the excessively fine granules which pervade the whole substance ; and in conformity with this it is usual to describe two kinds of white corpuscles—the finely granular and the coarsely granular. In addition to these there are others which appear almost perfectly clear and hyaline. As a rule, before the addition of reagents, no nucleus is visible in either variety, although, as will be afterwards seen, one or more is always present in each ; the nuclei are delicate, however, and readily obscured by the granules of the protoplasm. If the room is tolerably warm it may happen that the white corpuscles no longer preserve their rounded outline, but that from one side or another of a corpuscle a bud-like process extends itself, to be again retracted into the body of the corpuscle, spontaneous changes of form being thus effected which resemble those which are presented by the common fresh-water amœba, and are hence termed 'amœboid.' But in a cold preparation of human blood, like that under examination, these movements are seldom extensive, and do not serve to effect an actual change of place in the corpuscles such as we shall see to be the case in a preparation which is artificially warmed.

Further, there may generally be seen in a preparation of blood, in the clear interstices between the rouleaux of red corpuscles, a number of minute pale granules of a discoid shape (*blood-platelets*), which, if present in quantity, may be closely grouped together here and there into masses of various shapes and sizes, which the beginner is sometimes apt to mistake for white blood-corpuscles. But the objects in question have a much fainter aspect, and nothing resembling in nature the amœboid movements of the white blood-corpuscles is ever observed in them.

The masses in question are often of considerable size, many times larger than a pale blood-corpuscle. As Osler has shown, the particles which compose them are free in the circulating blood, and only run together when the blood is drawn.

Finally, a few delicate threads of fibrin may be observed stretching in different directions across the field of the microscope ; to see these distinctly a good lens is needed.

Blood on the warm stage.—In order properly to study the vital phenomena which are displayed by the white blood-corpuscles it is necessary, in the case of man and warm-blooded animals, to maintain the drop of blood under observation at or near the temperature of the body. For this purpose we employ what is known as a warm stage, of which there are several forms in use. The simplest consists merely of an oblong copper plate (fig. 35), two inches by one inch, from one side of which a rod of the same metal, four or five inches long, projects. This plate has a round aperture in the middle, half an inch in diameter, and is fastened to an ordinary slide by sealing-wax. The preparation is made as follows :—Take first a clean, long piece of covering-glass about 2 inches by 1 inch, which in this case is to be used instead of a slide, and on it make a preparation of blood obtained from the finger, as in the first preparation, carefully covering with an ordinary cover-glass. If there is now not enough blood to fill the space between the two glasses, add a little salt solution at one edge of the smaller cover-glass; but if, on the other hand, there is too much, soak up

FIG. 35

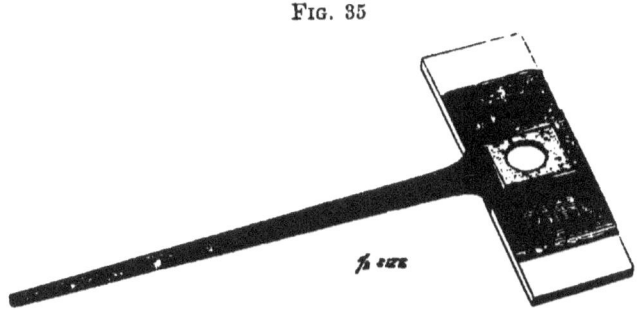

Simple warm stage, with preparation upon it enclosed between two cover-glasses

the excess with a small piece of blotting-paper. A very small camel-hair pencil which has been dipped in olive-oil is now to be drawn gently along each edge of the smaller glass : this will prevent evaporation from the edges, which would otherwise quickly ensue on warming the preparation. The glass slide which bears the copper plate having been clamped on to the microscope stage, with pieces of cloth underneath it to prevent the heat from being conducted away by the metal stage, the preparation thus made is placed upon the plate, and, having been brought in focus, one or more white corpuscles are selected for observation—a high magnifying power being used. The rod is now heated near its end by a small spirit-lamp, and the heat is conducted by the rod to the copper plate, and from this is transmitted to the preparation. A fragment of some fatty substance such as paraffin or a mixture of wax and cacao-butter, melting at about 35° C., is placed near one edge of the preparation, and another fragment, melting at about 40° C., near the opposite edge (fig. 36). The

Fig. 36

Simple warming apparatus, complete, shown in operation

lamp is now gradually approached along the rod until it arrives at a spot the heat transmitted from which is just sufficient to melt the one fragment but not the other, and it is then left burning at that spot; the preparation will then be maintained at a temperature approaching that of the blood.

Another way of mounting the blood is in the form of a thin 'hanging drop' on the under surface of the cover-glass, which is inverted over the hole in the copper plate. Before putting it down, moisten the glass slide which closes the hole below by breathing on to it. The edges of the cover-glass are then made air-tight by a layer of oil between it and the copper, and the layer of blood, which may, if desired, have been diluted with salt solution, now hangs within a shallow moist chamber. The advantage of adopting this method is that the preparation is freely exposed to the air of the moist chamber, and the white corpuscles retain their activity for a longer time unimpaired.

It will be seen that as the preparation begins to get warm the white corpuscles, which were perhaps previously rounded and inert, begin to throw out processes and exhibit amœboid movements, which become more marked as the temperature rises, so that by virtue of these an actual change of place from one part of the field to another may be effected. It is well in making this observation to select a single corpuscle and to sketch its outline and that of its more immediate surroundings at intervals of half a minute. As the corpuscles become spread out in creeping along the glass one or more nuclei may sometimes be seen indistinctly within them. Clear spaces or vacuoles are also to be seen in the protoplasm of some, whilst others are filled with granules, which stream with the movements of the corpuscles.

The red corpuscles in this preparation may be disregarded, for they show no trace of amœboid movement. The slight shaking movement which many of them exhibit is the molecular or Brownian movement common to all minute solid particles floating in a liquid.

After the observations recorded in the preceding paragraph are completed the action of an excess of heat may be observed; but it is better to use for this purpose a larger apparatus, in which the degree of heat can be measured by a thermometer. Such a one is shown in fig. 37. In this the preparation is placed upon the brass box a, which rests on the stage of the microscope, and is pierced in the centre by a tubular aperture to admit light to the object. The box is connected by india-rubber tubes with a hollow metal jacket, f, and the whole system thus constituted is completely filled to the exclusion of air with water previously boiled and cooled. The water is warmed at g by a small gas-flame, and rising through the tube c communicates its heat to the box a, the temperature of which is measured by a small thermometer, b, inserted through an obliquely placed tube quite into the central hole and immediately under the preparation. The cooled water from the

FIG. 37

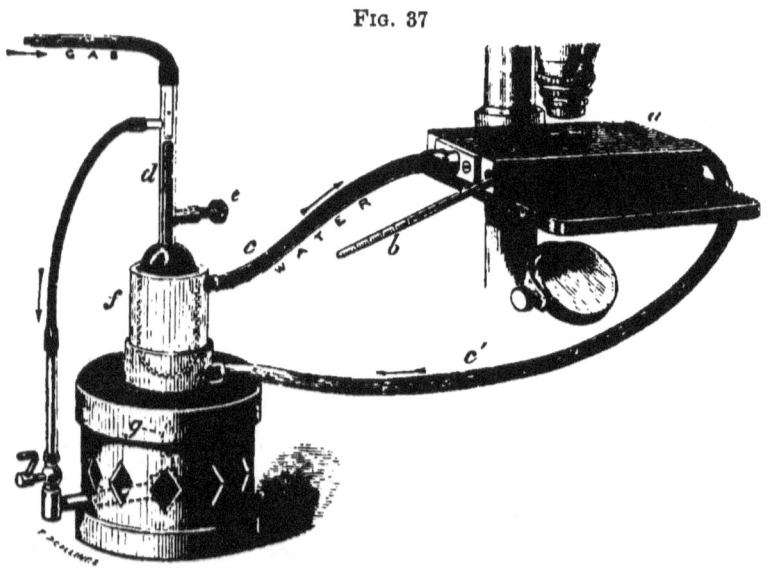

Apparatus for maintaining a constant temperature under the microscope [1]

stage descends down the tube c', to pass again round to the flame, and in this way the water constantly circulates. The bulbed tube d, filled with mercury, serves to regulate the flow of gas, so as to keep the temperature constant at any desired point. This is effected by turning the steel screw e, when this point, whatever it may be, is reached, so as to raise the mercury in the glass tube, and thus almost block up the lower end of a small steel or glass tube which is fixed into the upper

[1] See *Quarterly Journal of Microscopic Science*, 1875, vol. xiv.

end of the tube d. The gas passes through the small tube and then above the mercury and between the two tubes, to be conducted by the side piece h to the burner below; and it will be understood that if the temperature now rises higher in the reservoir f, which surrounds the mercury, this on being warmed will expand and tend to cut off more of the gas, and thus reduce the flame, on which the mercury will again contract, and the flame will rise in consequence, and so on. It is found that an equilibrium soon becomes established, and the temperature of the water and stage remains almost absolutely constant. To raise or lower the temperature all that is required is to screw out or in the screw e. The small included tube is pierced with a minute aperture, to allow a constant passage of gas, so as to prevent the flame from being extinguished in the event of the complete occlusion by the mercury of the lower end of the tube in question.

It will be found that up to and a little beyond the normal temperature of the blood the white corpuscles become more active in their movements, but on gradually warming the preparation still more, a point (50° C.) is reached at which they draw in their processes, become spherical, and show no longer any signs of vitality, the temperature being now sufficient to kill the corpuscles. The red corpuscles remain unaltered until a temperature of about 55° C. is reached, when they become altered in shape, and globular; soon they also begin to show alterations of outline and even to throw out beaded processes, but these changes are not due to any vital amœboid movements but to the action of heat in softening the envelope of the corpuscles. Eventually this envelope gives way altogether, and their colouring matter is then discharged and becomes dissolved out in the surrounding serum.

Another and a still more accurate method of observing the effect of varying degrees of heat upon a preparation is to enclose the whole microscope, except the upper end of the tube and the fine adjustment, in a metal box, which has a window in front for admission of light to the mirror, and a gas burner and regulator below for raising the temperature of the air within the box (fig. 38).

58   PRACTICAL HISTOLOGY

Action of Reagents upon the Blood.—The red corpuscles in the two preceding preparations appeared, even under the highest power, perfectly homogeneous and structure-

Fig. 88

Method of enclosing a microscope within a warm box

less; but it can be shown, by the application of reagents to the blood under the microscope, that they in reality consist of two separable parts—the envelope (formed of various chemical compounds, such as lecithin, cholesterin, nucleo-proteid), which is colourless and gives the shape to the whole corpuscle, and the *coloured* contents, which consist chiefly or wholly of a solution of a red crystallisable substance, hæmoglobin. The mode of application of reagents is as follows:—A drop of blood is got ready as in the first preparation, and whilst under observation a small drop of the reagent (which should as a rule be freshly prepared) is allowed to come into contact with the edge of the cover-glass. Some of the fluid flows under this and mixes with the drop of blood ; the current produced by it at first drives the corpuscles before it, but they soon become stationary, and then the part of the preparation should be selected for observation where the reagent is gradually diffusing itself amongst the corpuscles. In this way every stage in its action may readily be studied.

Action of water.—When a drop of distilled water is applied in the manner above described the first effect is seen to be that the red corpuscles begin to lose their discoid form, first one of their sides becoming bulged out so that they are cup-shaped, and then the other side, so that they are now completely globular, as may be seen when they roll over. Meanwhile the hæmoglobin is being dissolved out of the corpuscles by the water, so that they are soon quite colourless and hardly to be detected in the now reddish fluid. Some seem to offer greater resistance to the action of the water (and indeed of most reagents) and to retain their colouring matter longer than others.

The white corpuscles are also soon affected. They cease their amœboid movements and begin to swell up by imbibition of fluid, whilst at the same time with a high power the granules in their interior may be seen to exhibit the dancing movement which is characteristic of minute particles floating in liquid. Often the corpuscles present clear bulgings at their circumference, or their substance may appear to burst at one point and become diffused in the water. As they swell and become clearer, the nuclei generally come into view, and soon these also become swollen, and with the rest of the corpuscles eventually disintegrate, nothing being left but a few granules.

Water is thus proved to have a characteristic action upon the protoplasmic white corpuscles as well as upon the very easily alterable red discs, and this fact must be borne in mind in investigating the action of reagents or poisonous substances, both upon the blood-corpuscles and upon the tissues generally. If a reagent is to be employed in weak solution, therefore, it is well to dissolve it either in salt solution or in fresh serum [1] instead of water. Washing with water tissues which are subsequently to be submitted to microscopical examination is for a similar reason to be deprecated; but if a trace of bichromate of potash or of chromic acid, or a little common salt, be previously added to the water, its deleterious effect is in great measure obviated.

Action of acids.—To investigate the action of dilute acids it is best, as just explained, to mix the acid with salt solution instead of water: 1 part of glacial acetic acid to 200 of salt solution is an appropriate strength for the blood. The preparation is made in the usual way, and the drop allowed to run in at the edge of the cover-glass. The action of the weak acid upon the red corpuscles is seen to be quite like that of water: they are first rendered globular and then decolourised. Upon the white corpuscles it has a somewhat different action, for although the protoplasm of the corpuscle becomes partly swollen out into a clear spheroid, the nuclei are not swollen by the reagent, but are brought very distinctly into view, and remain usually at one side of the corpuscle, with a little granular matter precipitated around them.

Action of tannic acid.—The action of tannic acid upon the red corpuscles is peculiar and interesting. Like other acids it tends to cause the coloured part of the red corpuscle to pass out from the envelope; but as the coloured material is exuding it becomes coagulated by the astringent reagent, and in place

---

[1] The serum employed should be from the blood either of the same animal or of one belonging to the same species.

of being dissolved in the surrounding liquid, as after the action of acetic acid, it remains attached to the envelope as a small, bright, reddish projection.

In the first part of the reaction—viz. the rendering the corpuscles globular—tannic acid acts similarly to other weak acids.

The most convenient strength of solution to use is 1 per cent. ; it should be freshly prepared. It is well, moreover, first to mix the drop of blood upon which it is desired to test the action of the reagent with an equal amount of salt solution ; otherwise the tannic acid produces such a dense precipitate with the albumin of the serum that the view of the corpuscles is greatly obscured, and indeed the reagent with difficulty reaches them.

Action of alkalies.—A mixture of 1 part of caustic potash to 500 of salt solution may be used, a drop being added to the preparation in the usual way (p. 59). The reaction takes place very suddenly ; the corpuscles, both white and red, swell up as soon as the reagent reaches them, appear to burst, and then entirely disappear. The white are affected by a weaker solution than the red. This is apparent from the fact that, as the liquid slowly diffuses and mixes with the blood, the white may be seen to become destroyed in parts where the red are still unaffected.

Chloroform.—To observe the action of chloroform vapour upon blood a moist chamber is used. This is an apparatus for keeping a tissue or fluid under examination in its naturally moist condition, whilst at the same time allowing its surface to be exposed to air or to any desired gas or vapour. The simplest form of moist chamber is made of a small piece of soft putty or modelling-wax, which has been rolled out between the fingers into a round cord about 2 inches long and $\frac{1}{8}$ inch thick ; the ends of the cord are united so as to form a ring, and this is placed on the middle of a clean glass slide. A drop of water is put at the bottom of the chamber ; this is for the sake of keeping the atmosphere of the chamber moist ; but the object may be equally well effected by breathing into the space as it is being covered over. The object is prepared on the centre of a clean cover-glass, which is then inverted over the ring, so that the preparation is dependent into the chamber, and, whilst freely exposed to the air in this, is entirely protected from evaporation and may be readily examined through the cover-glass, to the under surface of which it remains adherent (fig. 39).

To investigate the action of chloroform vapour it is necessary to have some means of passing this into the moist chamber whilst the drop of blood is under observation. For this

FIG. 39

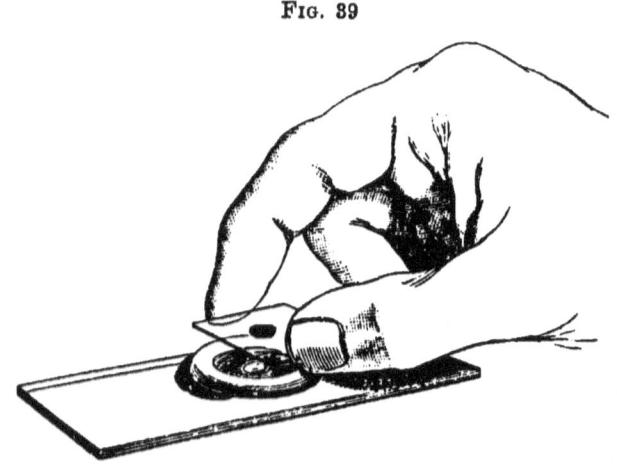

purpose a slide is employed (fig. 40), to which a piece of small glass tubing has previously been fixed by means of sealing-wax. This is done by heating the slide, dropping the sealing-wax upon it while hot, then heating the glass tube and laying it in the sealing-wax upon the slide. The ring of putty is so placed as to include the end of the tube, a small hole being made in the ring to afford an exit for the current of air containing the vapour. The slide is then clamped on to the microscope stage (as in fig. 46, p. 76), and the glass tube is connected by indiarubber tubing to a bottle containing a few drops of chloroform and furnished with a second tube, through which air can be blown.

Before the cover-glass is superposed the blood should first have been spread out upon it into a thin layer, so that the chloroform vapour may readily act upon all parts.

Everything being thus prepared, and some of the blood-corpuscles having been brought clearly under observation, air is blown gently into the bottle, and passing through it becomes charged with the vapour of chloroform, which is conveyed by the tube into the moist chamber, where it acts upon the layer of blood which is on the under surface of the cover-glass. After a short time it will be seen that the

FIG. 40

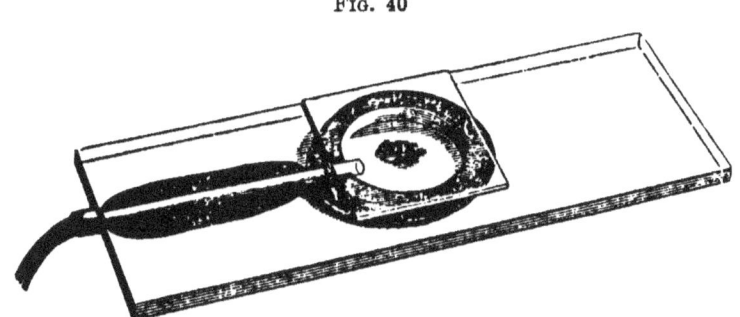

Chamber for passing a gas or vapour over a preparation under the microscope

amœboid movements of the pale corpuscles are arrested, and that the red corpuscles, as under the action of water and dilute acids, become globular; subsequently their hæmoglobin becomes dissolved and discharged out in the serum. It will further be observed, both in this and in the other preparations in which this change has taken place in the red corpuscles, that to the naked eye the blood has changed from scarlet to lake, and that whereas when the corpuscles were intact even a thin layer of blood presented a somewhat opaque appearance, it is now completely transparent. Hence we may infer that the opacity of the unaltered blood is due to the presence of the red particles.

Enumeration of blood-corpuscles.—The following are necessary for counting the blood-corpuscles. (*a*.) A diluting solution. Either that recommended by Hayem may be used, viz. :—

| Distilled water | . 200 cc. | Common salt | . 1 grm. |
| Sulphate of soda | . 5 grm. | Corrosive sublimate | . 0·5 grm. |

or Sherrington's fluid, viz. :

        Distilled water . . . . . 300 cc.
        Common salt . . . . . 1·2 grm.
        Neutral pot. oxalate . . . . 1·2 grm.

to which eosin may be added if desired.

In the latter fluid, if means be taken to prevent evaporation, the white corpuscles remain living and amœboid for days, and the granules become stained *in vivo*.

(*b.*) A measuring pipette holding 10 cubic millimetres, and about 1 centimetre long. This is made out of a piece of thick-walled capillary tube, and is ground and polished to a blunt point at one end. It has a flat narrow band of German silver attached to its middle, to serve as a handle (fig. 42, *a*).

(*c.*) A cylindrical mixing vessel marked to be filled up to either 50 or 100 times the capacity of the pipette.

Fig. 41

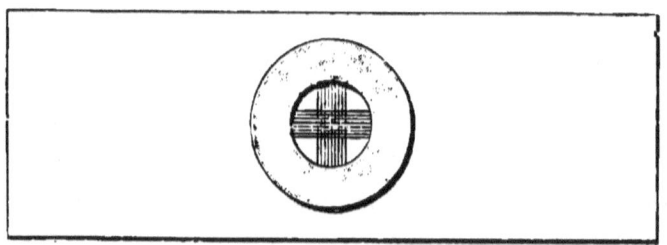

Slide ruled in squares of ·1 mm. each, for enumerating blood corpuscles under the microscope

(*d.*) A counting slide, marked at its centre by a diamond with lines forming squares of 0·1 millimetre, and having a ground-glass ring cemented to it which is exactly either 0·1 or 0·2 millimetre thick (Gowers). The slide may be fitted with springs to press the cover-glass down firmly on the ring.

(e.) A small glass stirrer, needles, worsted, cover-glasses, and a dropping tube (fig. 42, b).

Proceed as follows :—See that all the apparatus is clean and dry. (The capillary pipettes can be readily dried by passing worsted through them with a needle.) Prick the finger, squeezing it slightly to force the blood out more quickly, and when a large enough drop has exuded, place the point of a capillary pipette in it, and allow the pipette to fill completely with blood, which it will easily do by virtue of its capillarity. Wipe off with the finger any blood which may have got on the outside of the capillary. Hold the capillary over the mixing vessel, and from the dropping tube let some of the diluting solution flow through the capillary, thus completely washing out the blood, until the mixer is filled up to the 100 mark. The blood is thereby diluted to form 1 per cent. of the mixture. If it be desired to form a 2 per cent. mixture, the vessel must only be filled to the 50 mark. This is a better degree of dilution when the white corpuscles are to be enumerated as well as the red. Now stir the mixture and place a drop of it upon the counting slide, so that on placing the cover-glass upon the ring it (the cover-glass) comes in contact with the top of the drop. Assuming the ground-glass ring to be 0·1 millimetre thick, the mixture will then form a layer 0·1 millimetre thick between the cover-glass and slide, and the liquid above each square will be 0·1 millimetre cube. In a few minutes the blood-corpuscles will have sunk to the bottom, and will rest on these squares, where they may be counted. The number on each square represents the number in 0·001 ($\frac{1}{1000}$th) cubic millimetre of the mixture. To arrive at the number in 1 cubic millimetre it is best to count the number on ten squares and multiply by 100. The product multiplied by 100 or by 50 (according as a 1 per cent. or 2 per cent. mixture was used) gives the number in a cubic millimetre of *undiluted blood*. Of course the number on ten squares may be at once multiplied by 10,000 or by 5,000, as the case may be. It will be found that in normal human blood (male) there are on an average 5,000,000 red corpuscles to the cubic millimetre. For the white corpuscles it is desirable to count the number on a much larger number of squares.

Oliver's method.—A much more ready and at least equally accurate method of estimating the number of coloured corpuscles in blood is that devised by Dr. G. Oliver. A small quantity of blood is taken up as before into a short pipette (fig. 42, a), and at once washed out of this by the dropping tube, b, into a graduated flattened test tube (c), with Hayem's diluting mixture (which must not contain colouring matter). The graduations of the tube are so adjusted to the capacity of the pipette that with normal blood (assumed to contain 5,000,000 red corpuscles to the cubic millimetre) the light of a small wax candle placed at a distance of three yards from the eye in a dark room is just transmitted as a fine bright line when looked at through the tube held edgeways between the fingers (d) and filled up to the 100 mark of the graduation. If there are fewer corpuscles than the normal, less of the diluting solution is required for the light to be transmitted; if more than normal, more of the solution must be added. The tube is graduated above and below the 100 mark so as to indicate in percentages every decrease or increase of corpuscles per cubic millimetre, as compared with the normal standard of 100 per cent. By this means an accurate result can be obtained in two or three minutes, whereas by the most expeditious observer an actual enumeration will take from 10 to 15 minutes.[1]

Examination of frog's or newt's blood.—The simplest way to obtain frog's blood for examination is to cut off the tip of one of the digits, having previously wiped it dry with a cloth, and to collect upon a clean cover-glass the small drop of reddish fluid which exudes. The glass is then inverted upon a slide and the drop is examined. The blood so obtained is mixed with lymph, and the corpuscles are consequently less

[1] Oliver's apparatus is supplied by the Tintometer Company, 6 Farringdon Avenue, E.C.

FIG. 42

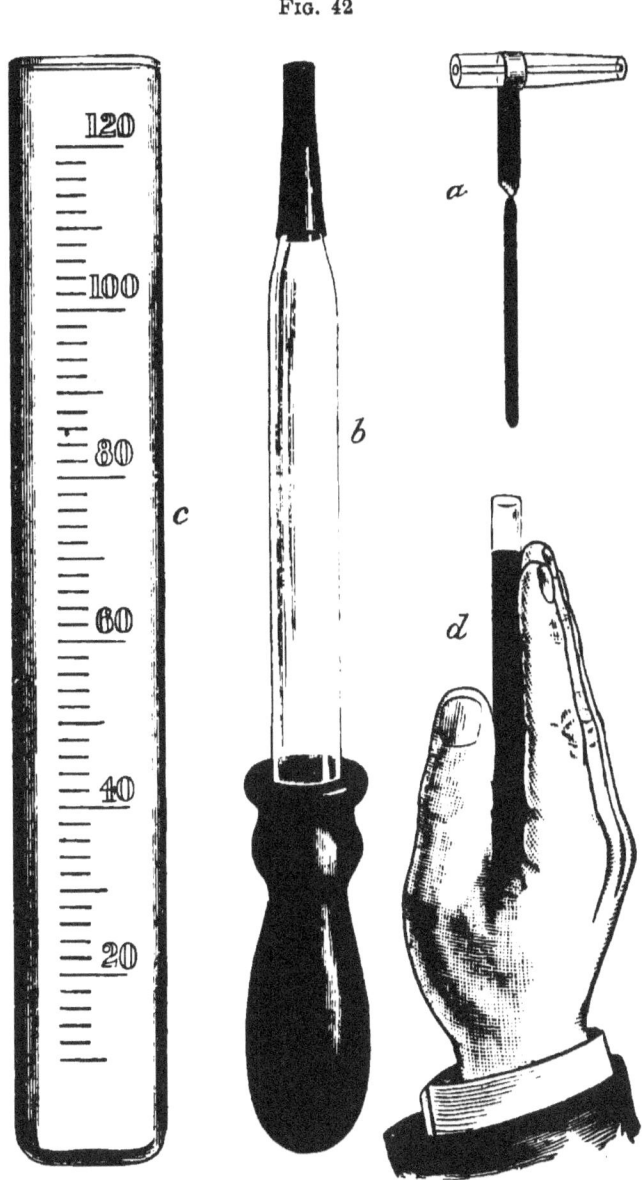

Oliver's apparatus for rapidly estimating the number of corpuscles in a sample of blood by means of the opacity method

crowded and better adapted for observation than when the blood is undiluted. For newt's blood the end of the tail is cut off, and a drop of blood similarly obtained. To procure blood unmixed with lymph the animal should be pithed, and laid upon its back. The heart is then exposed and snipped with scissors, and a small drop of the blood which exudes is taken up with a glass rod, transferred to a slide and covered. But before placing the cover-glass down in the usual way a small length ($\frac{1}{4}$ inch) of a delicate hair should be placed in the drop, so that when the cover-glass settles down, the corpuscles, here comparatively large, may not be crushed by its weight.

When the preparation obtained from either of these sources is examined it will at once be seen that the coloured corpuscles are larger and fewer in number than in human blood, and do not tend to form rouleaux, although they adhere together in a less regular fashion; that they have an elliptical outline when lying flat, but when seen edgeways look quite narrow and pointed at the ends, with a slight and gradual bulging at either side; so that, although disc-shaped, like the mammalian blood-corpuscle, so far from being biconcave, they are biconvex. The bulging is due to the presence in the middle of the corpuscle of another part besides the envelope and the coloured contents found in the mammalian disc. This, the *nucleus*, can readily be made out in most of the corpuscles as a somewhat elongated, colourless, and slightly granular elliptical body, about a third the length of the corpuscle, and often lying not quite in the middle but somewhat excentrically. Occasionally the nucleus is seen to be round; but this is an accidental change, and may be brought about by mechanical injury. Indeed, if the precautions above recommended for avoiding pressure are not taken, a large number of the corpuscles become injured, so much so as even to be ruptured and destroyed altogether, in which case the rounded nuclei are liberated, and may be mistaken by the beginner for pale blood-corpuscles.

With regard to the white corpuscles, it will be observed

that they are more numerous and larger than in human blood. Moreover, they soon begin, even in the cold, to exhibit very distinct amœboid movements; and, on account of the greater size of the corpuscles, both these and the other phenomena exhibited by them are much more striking, and these corpuscles are therefore much better suited for observation than those of mammals, which in other respects they closely resemble. The distinction between the finely- and the coarsely-granular corpuscle is met with again here; but there is also sometimes seen a third kind of corpuscle, fusiform or oat-shaped and devoid of amœboid properties.

Feeding of the white corpuscles.—The white corpuscles of the blood exhibit a strong tendency to take into their interior any small particles of solid material which may happen to be in their neighbourhood. This tendency they have in common with all amœboid organisms, and it is nowhere better seen than in the case of the amœba itself. Because of their greater size, number, and activity, it is better to take the newt's corpuscles for this experiment than those of the mammal.

Any substance consisting of fine insoluble particles, such as vermilion or Chinese ink, may be rubbed up with salt solution and used for this experiment, but the most convenient material is yeast. A very small fragment of German yeast is rubbed up with a little salt solution so that the solution is faintly milky. A small drop of this mixture is then placed on a slide, and a drop of blood is added to and well mixed up with it. The preparation is then covered, and the edges of the cover-glass are painted with a film of oil to preclude evaporation. If the white corpuscles are now observed, it will be seen that certain of them gradually take into their substance the yeast particles with which they happen to come in contact in their amœboid movements. On careful watching it may be made out that the process of inception commences by the throwing out of processes which surround the particle to be taken in, and meet and coalesce beyond it; once included in this manner, the granule afterwards becomes gradually carried, presumably by the movements of the protoplasm, more towards the centre. If, after observing the preparation in this way for half an hour, it be laid aside for two or three hours, it will be found at the expiration of that time that many of the white corpuscles have taken in a large number of the yeast particles, for they do not discharge their cargo, but carry it about in their movements from place to place.

It will be found that it is the large corpuscles of the finely granuled kind which exhibit this tendency to the most marked extent, the corpuscles with coarse granules, although amœboid, are not phagocytes.

Migration of white corpuscles.—The process of migration can be quite easily seen in progress in transparent parts of animals, and will be afterwards studied when the methods of observing the circulation of the blood are described. But for exhibiting the active migration of the white blood-corpuscles nothing is more striking than the examination of a capillary glass tube in which frog's blood has been collected and has coagulated. A high power is to be employed, and the capillary tube must therefore be very fine ; and in order that the wall of the tube should be as thin as possible, it must be drawn out from a piece of large and thin tubing. The capillary tube is filled with frog's or newt's blood, except near the ends ; these are then sealed by holding them successively in the flame for a second ; the tube is then placed in a drop of cedar oil or thick Canada balsam on a slide, covered with a thin glass, and at once examined. The object of the essential oil is to correct in some measure by its high refracting power the effect upon the light of the cylindrical glass tube. After a few minutes the clot is seen to be getting smaller, and a layer of clear serum collects between it and the glass ; the quantity of this gradually increases, and soon portions of white corpuscles begin to project from the surface of the clot. These protrude more and more, and others make their appearance, and all begin to throw out numerous amœboid processes, which are actively advanced and retracted. By aid of these the corpuscles gradually emerge from the shrinking clot, and eventually become free in the surrounding serum (fig. 43).

Influence of warmth on white corpuscles of Amphibia.—
The action of gentle warmth in accelerating the movements of the pale corpuscles of the frog or newt may be investigated with the same apparatus as was used for the observation of mammalian blood at the temperature of the body. But it will

FIG. 43

White corpuscles of frog's blood migrating from clot. Highly magnified. The clot has shrunk considerably from the sides of the capillary tube

be found that if the temperature be allowed to rise so high as 38° C. the movements of the corpuscles of these cold-blooded animals will soon permanently cease, the corpuscles being killed (heat-rigor). So that unless this result is desired, a paraffin of lower melting-point must be employed to indicate the temperature limit which is not to be exceeded.

Action of electric shocks.—For this a slide must be specially prepared by cementing to its upper surface with shellac varnish two slips of gold-leaf or tinfoil, with pointed ends which almost meet in the middle of the slide (fig. 44, A). A drop of blood is put here, the cover-glass is placed over it, and the portion of blood which lies between the points is brought under observation. Or a moist chamber may be employed,

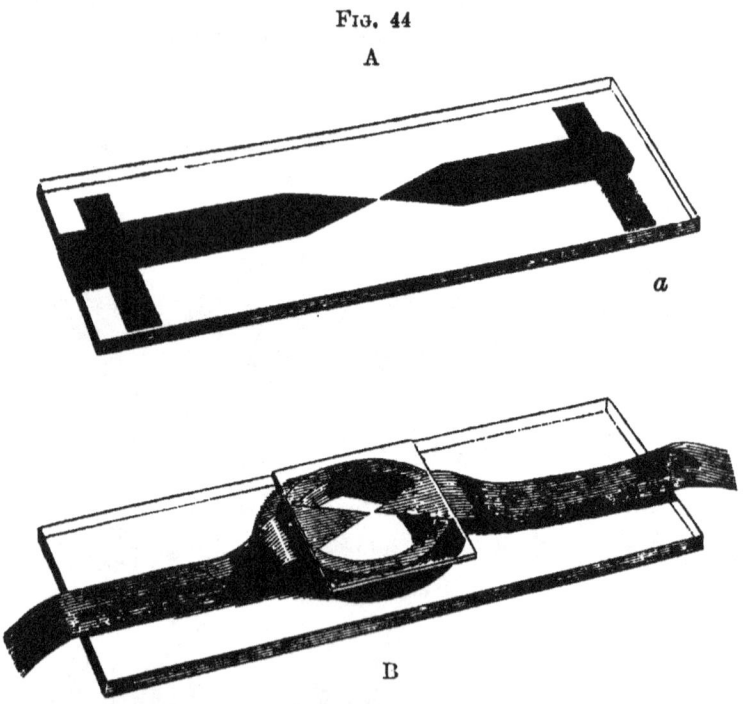

Fig. 44

Glass slide, with two strips of tin-foil, one of which passes round to the under surface, where it rests on the brass stage of the microscope; the other strip is isolated from the stage, and may be connected to the outer coating of a Leyden jar, the charge of which is made to pass between the points by connecting the knob of the jar with the brass-work of the microscope. Opposite a, a small piece of the foil is fixed to the under surface of the slide, so that this end shall be level with the other

the cover-glass used having previously had two strips of tinfoil cemented to it (fig. 44, B). The drop of blood being spread out in a thin layer between their points, is quickly inverted over the ring of putty and brought under observation. The tinfoil slips are kept isolated from the brasswork of the microscope, and are so arranged that the charge of a small Leyden jar, or an induced current of electricity, can be passed through them at any moment (fig. 45). One or more amœboid white corpuscles which happen to be in the path which the spark must take in traversing the interval between the points are kept in view, and the spark is then allowed to pass. The white corpuscles immediately cease moving, withdraw their

FIG. 45

Apparatus for passing electric shocks through a drop of blood, which is to be examined in a moist chamber. The tin-foil slips are cemented near their points to the under surface of the cover-glass, and their free ends are clamped to isolated metal supports, connected by wires to an induction-coil. The tin-foil slips are isolated from the brass stage of the microscope by the glass slide on which they rest

processes, and become rounded in shape; in fact, undergo general contraction. But if only one slight shock be given they soon recover and resume their movements, although these are often somewhat altered in character. The red corpuscles are but slightly if at all affected; but if a succession of shocks are transmitted from an electric machine or an induction-coil, electrolytic action is set up in the fluid, bubbles of gas are developed, the effects respectively of acids and alkalies are set up in the neighbourhood of the tinfoil points, and the red corpuscles undergo changes brought about by these.

Presence of glycogen.—Many of the white corpuscles contain a certain amount of glycogen, either in distinct granules or in a more diffused form. This substance becomes stained of a reddish mahogany colour by solution of iodine, and may thus be readily detected, both here and elsewhere. The solution to be used is made by dissolving 1 gramme of iodine in 100 cc. of water which contains 2 grammes of iodide of potassium in solution.

The preparation, preferably of frog's or newt's blood, is made in the usual way, and the iodine solution added at the side of the cover-glass. The red corpuscles are stained of an intense yellow, but are otherwise little altered, except that the nucleus, which remains unstained, becomes globular, and bulges out at either surface of the corpuscle. The white corpuscles are instantly arrested in their movements and killed, preserving exactly the form which they exhibited when reached by the iodine solution. Being of less specific gravity than the latter, they tend to float on it ; and, if the layer of fluid is thick, must be sought by focussing upwards in the stratum immediately under the cover-glass. The main substance of the corpuscle is uniformly stained of a deep yellow, but many contain groups of more darkly stained granules, and from others are seen to exude after a time pellucid drops of varying size, which become tinted of a pale port wine colour, and no doubt contain glycogen.

Action of reagents—water, acids, &c.—upon the frog's blood-corpuscles.—The action of reagents upon the white blood-corpuscle of the frog is exactly the same as upon the mammalian white corpuscle ; upon the red corpuscle it is in the main similar, but in some cases a little different, the differences partly depending upon the presence of the nucleus. Thus *water* passing through the envelope, after causing the corpuscle to swell up and both it and the nucleus to become spheroidal, extracts the coloured contents of the corpuscle, which usually resumes its oval shape after the hæmoglobin has escaped ; dilute *acetic acid* brings the nucleus strongly into view, and decolourises the rest of the corpuscle ; *tannic acid* causes the coloured part to be exuded from the corpuscle, to the envelope of which it generally remains attached as an irregular curdled mass (sometimes the coloured part is precipitated around the nucleus, and then the two may be ejected from the stroma together) ; *chloroform* vapour causes the red corpuscles to become decolourised, and arrests the movements of the white corpuscles ; but these, if not acted on for too long a time or by too strong a mixture of chloroform and air, are resumed on replacing the vapour by pure air. All these reagents are applied in precisely the same manner as with mammalian blood, to the description of which the reader is referred.

The observations may be made with as great or with greater advantage upon the blood-corpuscles of the newt, which are larger than those of the frog. The blood should preferably be obtained directly from the heart, and not by merely snipping a piece off the tail and collecting the drop which exudes, for in this case it is very apt to become mixed with the acrid secretion from the cutaneous glands. There are in addition two reactions for which newt's blood is particularly well adapted, viz. the actions respectively of boracic acid and carbonic acid upon the red corpuscles.

Action of boracic acid.—The boracic acid is used in solution in water (2 per cent.), and the preparation of blood having been made in the usual way, with or without addition of salt solution, a drop of the boracic acid solution is placed at the edge of the cover-glass and allowed slowly to mingle with the blood. If the first stages of the reaction are fortunately observed, it will be seen that the coloured part of the corpuscle is becoming collected towards the centre of the corpuscle and accumulated around the nucleus, often remaining, however, at first adherent here and there to the circumference of the corpuscle, and shrinking away at the intermediate points, so as to present somewhat of a stellate figure.[1] But soon it is entirely withdrawn and collected around the nucleus, which has become rounded, and is nearly concealed by the colouring matter. The corpuscle, now decolourised, has also in many cases become circular, and the coloured nucleus is generally shifted to one side, and eventually altogether extruded. Another even more ready method of obtaining this reaction is to drop the blood directly into the solution of boracic acid.

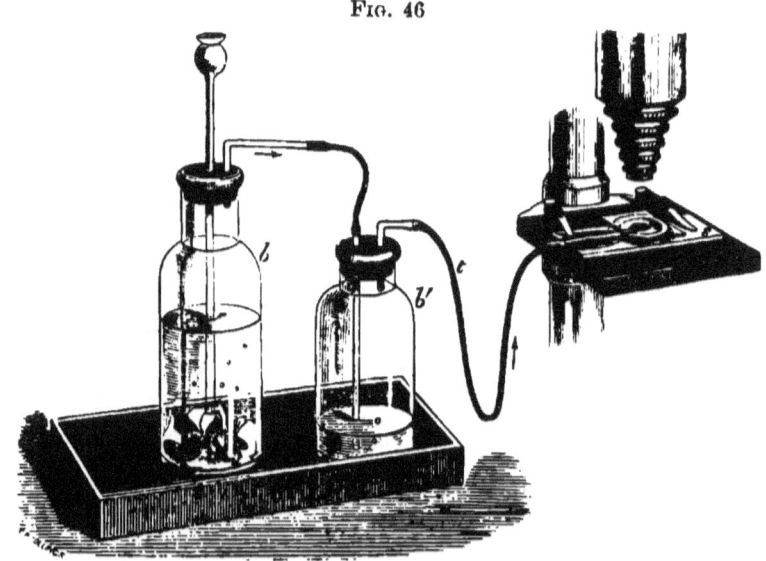

FIG. 46

Apparatus for passing carbonic acid gas over a preparation under the microscope

b, bottle containing marble and hydrochloric acid ; *b'*, wash-bottle ; *t*, india-rubber tube conducting the gas to the stage, *s*

Action of carbonic acid.—To investigate the action of carbonic acid gas the blood must be prepared in a moist chamber, like that used for chloroform vapour (fig. 40). If the preparation is very quickly made the nucleus in many of

[1] Water and various other reagents may, in the first instance, have a somewhat similar effect.

the red corpuscles cannot at first be distinguished, for in the entirely unaltered state it possesses as nearly as possible the same index of refraction as the rest of the corpuscle. But when carbonic acid gas, generated in a suitable apparatus (fig. 46, b, b'), is allowed to pass into the moist chamber a fine cloudiness or precipitate occurs around the nucleus, and the outline of this is brought into view, whereas if the carbonic acid is speedily replaced by air, which may be effected by disconnecting the tube t from the wash-bottle b' and drawing air through it by the mouth, the precipitate is re-dissolved, and the nucleus is again made to disappear.

It is of advantage in performing this experiment to add a trace of moisture to the blood before the observation; this may be most conveniently done by breathing two or three times on the preparation before placing it over the ring of putty.

Crystals obtainable from the colouring matter of blood. hæmoglobin crystals.—The hæmoglobin or colouring matter of the blood may be obtained in definite crystals, but the form of the crystals varies in different animals. It is difficult to induce the crystallisation in human blood; and to obtain the crystals readily it is best to employ the blood of the dog, rat, guinea-pig, or squirrel.

If a drop of rat's blood is mixed with an equal amount of distilled water, the hæmoglobin becomes extracted from the corpuscles; until this is done no crystallisation takes place. As the excess of water begins to evaporate small needle-shaped crystals of hæmoglobin appear, either singly or in bunches, which become gradually larger until they may attain a very considerable size. They are mounted by covering them in thick Canada balsam.

Another method is the following :—

The animal is bled, and the blood as it flows from the divided vessels is vigorously stirred with a bundle of wires, to remove the fibrin. A small quantity of the whipped blood is then mixed with about one-third its volume of water, and a drop of chloroform [1] being added, the mixture is thoroughly shaken up for a minute or two. This has the effect of discharging the hæmoglobin from the corpuscles into the surrounding fluid. A small drop is now placed upon a slide and left exposed to the air for a few minutes. It becomes thickened by evaporation and dried at the edges, and crystals, tetrahedral in form from the guinea-pig, hexagonal from the squirrel, and rhombic needles from other animals, may be detected in it with a low power of the microscope. The drop may then be surrounded with thick Canada balsam, covered and examined with a higher power. The crystals increase in size for a time, and new ones continue to be formed.

Hæmoglobin crystals will not keep unaltered for an indefinite time.

Hæmin crystals.—The name 'hæmin' has been given to certain very characteristic crystals which are formed at the expense of the colouring matter of the blood, and the production of which is a trustworthy test of the presence of blood, although yielding no indication of the kind of animal from which the blood has been obtained. To see them a very small quantity of blood obtained from the finger or elsewhere is smeared upon a glass slide and allowed to dry. A cover-glass is then placed over it, and a drop of glacial acetic acid is applied from a pipette to the edge of the cover-glass and allowed to run under by capillary attraction. The slide is then held by one end in the fingers, and the middle is gently warmed over a small flame. As soon as bubbles begin to appear in the fluid the warmth is discontinued, and the preparation is examined with a high power. If no crystals appear as the slide cools a little more acid is added, to replace that lost by evaporation, and the slide is warmed as before, and on cooling again examined. It will be found that almost all over the preparation reddish-brown short prismatic crystals, disposed singly or in groups, will have made their appearance;

[1] If nitrite of amyl is used instead of chloroform, crystals of *methæmoglobin* become formed in place of hæmoglobin (Halliburton).

most of them are very minute, but they may be obtained of considerable size by rewarming the preparation with glacial acetic acid once or twice.

The presence of a chloride is necessary for the formation of these crystals; in the case of recent blood the chlorides which it naturally contains are sufficient for the purpose; but if it were an old blood-stain which one had to deal with, in which the chlorides may have been washed away, it would be previously requisite to mix a minute quantity of common salt with the stain to be tested in order to supply the deficiency.

Methods of fixing and permanently preserving blood-corpuscles.—No reagent is perfectly satisfactory for fixing the red blood-corpuscles, but osmic acid and corrosive sublimate are the best for the purpose. To apply the former, a drop of blood is directly mixed as it flows from the pricked finger or from the vessels with an excess of 1 per cent. osmic acid, which may contain eosin in solution. The mixture is allowed to stand for an hour protected from evaporation, and some of it is then mounted in dilute glycerine.

Another and on the whole better method of applying osmic acid is in the form of vapour. A very thin drop of blood is placed upon a clean cover-glass and instantly laid face downwards covering the mouth of a small bottle containing 1 per cent. solution of osmic acid. The vapour of the acid fixes the blood-corpuscles, and after the preparation has remained for a few minutes in this position, the cover-glass with the layer of blood is removed and placed on a slide in a small drop of glycerine and water, coloured by eosin. If the blood is too much clumped together, the cover-glass may be gently tapped to separate the corpuscles.

Corrosive sublimate may also be employed to fix the blood-corpuscles, in the form of Pacini's fluid (mercuric perchloride 1 gramme, pure sodic chloride 1 gramme, distilled water 200 cubic centimetres) or Hayem's fluid (see p. 64). The blood is mixed with many times its volume of the fluid as it flows from the pricked finger or from the vessels, and is left for from 15 minutes to several hours for the corpuscles to settle. The fluid is then decanted off and replaced by rectified spirit containing a little tincture of iodine, and afterwards by 96 p.c. alcohol. The corpuscles are stained by dilute solution of methylene-blue or by eosin, or by eosin and methylene-blue in succession.

Blood thus treated may be preserved either in dilute glycerine or in Canada balsam. The methylene-blue and eosin stain respectively the baso-phil and oxy-phil granules of the leucocytes. Eosin also stains the coloured corpuscles. A better plan to stain the granules of the leucocytes is the following :—Make a preparation of blood between two cover-glasses, allow it to stand for a minute or two, then separate the cover-glasses and let the blood upon them dry quickly either at the ordinary temperature of the air or by holding them for an instant over a flame. If it is desired to stain the coloured as well as the colourless corpuscles, the cover-glasses are either heated for an hour or more to 120° C., or placed for an hour in a mixture of equal parts of absolute alcohol and ether, to fix the hæmoglobin ; if not, the staining may be proceeded with at once. For this purpose the cover-glass preparation is placed in an aqueous solution of the stain which it is desired to employ, *e.g.* eosin or methyl-green, or any of those enumerated under aniline dyes (p. 21), or in Ehrlich-Biondi mixture, preferably for some hours, but the process may be accelerated by warming the solution. The cover-glass is then washed with water, with acidified spirit, and finally with absolute alcohol, until the film is nearly colourless, when it is passed through xylol, to be mounted with xylol balsam. The staining may sometimes be advantageously effected with alcoholic solutions of the dyes.[1]

To fix the white blood-corpuscles in their amœboid condition the following method may be recommended. Mix

---

[1] See Kanthack and Hardy, *Journ. Physiol.* vol. xvii., and Hardy and Westbrook, *Journ. Physiol.* vol. xviii.

newt's blood or lymph taken from the peritoneal cavity with a little normal saline solution, cover, and gently irrigate the preparation with normal saline so as to wash away most of the plasma: the white corpuscles are not washed away, because they tend to adhere to the glass. Put the preparation aside for ten minutes, by which time most of the white corpuscles will be actively amœboid. Then allow a jet of steam from the spout of a kettle, or from a glass tube fitted to a flask of boiling water, to play for one or two seconds upon the cover-glass, holding the preparation close up to the spout. The heat instantly fixes the corpuscles; and the preparation is then irrigated with dilute alcohol, and afterwards with dilute hæmatoxylin or with one of the aniline dyes.

After the corpuscles are stained the excess of colouring solution is washed away by a little dilute alcohol (50 per cent.); this, in its turn, is replaced by absolute alcohol, and this by bergamot oil; finally xylol balsam is passed under the cover-glass.

A very simple and ready method of obtaining a permanent stained preparation of frog's blood-corpuscles is the following, which is a modification of that given by Stirling:—Mix the blood directly with a quantity of Flemming's fluid (p. 16). After half an hour decant off the Flemming's fluid and rinse with water, decanting this off also. Then add picrocarmine solution, and allow this to stand on the blood-corpuscles for some hours or even days. Decant off the picrocarmine, and mix the stained residue of corpuscles with a little glycerine jelly rendered fluid by heat. A drop of this can be taken and mounted at any time, for it will keep indefinitely. The nuclei of the corpuscles are stained red, and the bodies of the corpuscles yellow. This method of preserving specimens in glycerine jelly is very suitable for class purposes, and may be applied to other preparations of isolated elements, such as macerated epithelia, nerve- and neuroglia-cells, muscle-fibres, and the like.

CHAPTER II

THE EPITHELIAL TISSUES

THE epithelial tissues are studied with regard both to the structure and form of the individual elements, and the relations these bear to one another and to the membranes they cover. The latter class of observation can only be properly made by the study of sections of the various organs and parts where epithelium is found, and will therefore be left until the method of making these is explained. The modes of isolating and studying the individual cells will, however, be best described in this place.

Scaly epithelium. Superficial layers.—If a little of the saliva which moistens the inside of the cheek be gently scraped off with a small spatula or with the finger-nail, a number of the superficial cells of the thick stratified epithelium which is here met with will be brought away with it. The material thus obtained is placed upon a slide and a cover-glass put over. On examining the preparation numerous flattened scaly epithelium cells will be seen, either entirely isolated or in little patches, the cells in a patch being connected together, with their edges overlapping. The cells are of considerable size, each with a nucleus near its middle, small in comparison with the size of the cell; and the substance of the cell, although clear, yet contains a number of scattered granules. Moreover, lines may often be seen running in various directions over the surface; these are for the most part due to creases of its substance caused by the pressure of adjoining cells. Some of the cells may be seen edgeways, and then, being flattened, will appear narrow and linear; but on touching the cover-glass with a bristle their true form will be apparent as they turn over.

In addition to such cells as these a certain number of much smaller rounded cells may generally be seen in the saliva, which somewhat resemble the white corpuscles of the blood, and, like them, frequently exhibit amœboid movements. They are, in fact, lymph-corpuscles which have come from the mucous membrane covering the back of the tongue and the tonsils, which is very rich in these cells. The saliva being a watery fluid they are swollen out by it, and with a good microscope it may be observed that the granules in the interior of the corpuscles exhibit the Brownian molecular movement, a phenomenon which, it will be remembered, was exhibited by the white corpuscles of the blood as a first result of the imbibition of water.

There are several layers of the above-described large flattened epithelial cells in the epithelium of the mouth. Below them are other cells, smaller and of a more spheroidal or polyhedral shape, and joined to one another by fibres which pass across fine intercellular spaces. To obtain these deeper cells isolated, it is necessary to macerate a piece of any membrane which is covered by a stratified epithelium in some fluid which, while tending to dissolve the intermediate substance which cements the cells together, may preserve their natural form and, at the same time, prevent putrefaction from appearing in the tissue which is undergoing maceration. The best liquid for this purpose is a mixture of alcohol and water (one part alcohol to two parts water). The portion of tissue must be small, and the quantity of liquid used comparatively large. A piece of the mucous membrane of the mouth, pharynx, or gullet of any mammal may be used; it will require at least two or three weeks' maceration (for in this kind of epithelium the cells are very closely united), and the solution should be changed every third day. At the expiration of the time stated a small portion of the epithelium is scraped off with the point of a knife, and placed in a drop of water upon a slide. It is then broken up as finely as possible with a pair of mounted needles, a piece of hair is cut off and placed in the drop, and the cover-glass is superposed, after which the pieces of tissue can be still further broken up by tapping on the cover-glass. By far the majority of the cells seen are the superficial ones already described; but others will be found which are less flattened and smaller in diameter, and have many of them an irregular toothed margin, their surface also, as may be seen by altering the fine adjustment of the microscope, having on it linear or punctated markings. These are the so-called cells with ridges and spines, the spiny appearance being caused by the fibres which passed from cell to cell, and which have become broken across in the process of separation. The cells will be again studied in sections of the skin, as well as in sections of mucous membranes which are covered with stratified epithelium.

Horny layer of epidermis.—If a very small shred of the superficial part of the epidermis is taken from any part, the palm of the hand, for instance, and examined in water under the microscope, no indications of cellular structure are visible—nothing but an irregular confused mass is to be seen. Remove the cover-glass and place the shred of epidermis in a drop of liquor potassæ. It will soon swell up and become soft. When this is the case return it to the water, and, after breaking it up as finely as possible with needles, cover and examine. Numerous spheroidal cells are now seen loose in the preparation, with a distinct contour, as if enclosed by a membrane, but without a nucleus. They are, in fact, the scaly cells, which have become swollen out by imbibition of water, and, at the same time, in consequence of this swelling of their surfaces, correspondingly diminished in width.

The same result may be obtained with the cells which form the nails; and also with the flattened cells from the mucous membrane of the mouth.

Columnar epithelium.—This is most characteristically seen and is best studied as met with in the intestinal canal.

It must be taken from an animal (amphibian or mammal) quite recently killed, as it rapidly undergoes destructive changes if left after death in contact with the intestinal contents. When the intestine is cut across at any part in a recently killed animal the cut edges curl outwards, and a little of the mucous membrane is thus exposed. Two very small portions may be snipped off this. One of these is placed for a few hours in a few drops of a 1 per cent. solution of osmic acid, and is then transferred to water, whilst the other is immersed in Flemming's solution diluted with 100 times its bulk of salt solution, or in dilute chromic acid solution, made by dissolving 1 part of chromic acid in 2,000 of salt solution.[1]

These two portions may be put aside for the present in their respective fluids, whilst a preparation of the epithelium in the fresh condition is made and examined. For this purpose slit open a piece of the intestine, wash away the mucus and intestinal contents by allowing a little serum or normal salt solution to drop upon the inner surface, and then, with the end of a clean scalpel, gently scrape the washed surface, and transfer what is brought away on the scalpel to a drop of fresh serum upon a clean glass slide, and cover the preparation, averting the pressure of the cover-glass by means of a piece of hair. On examining with a high power the specimen so obtained numerous columnar epithelium cells will be seen, some separate, others in groups. (It will be found advantageous in examining the object to moderate the illumination by the employment of one of the smaller holes of the diaphragm : in this way the somewhat indistinct outlines of the tissue-elements are rendered plainer, and any details of structure can generally be more readily made out ; this is the case, indeed, with all unstained preparations.) In the separated cells the conical form

---

[1] One-third alcohol may also be employed, and this serves to separate the cells still better than the other fluids, but their structure is not so well preserved by it. The isolated cells can be stained with picrocarmine or carmalum and preserved in glycerine jelly in the manner recommended for the amphibian blood-corpuscles (p. 81).

may be seen, the cell tapering at one end to a point, or terminating in a rounded or flattened extremity. The general substance of the cell has a faintly granular appearance in the fresh condition, without a distinct outline, except at the larger end, which is bounded by a strongly refracting thickened margin, in which a few faint striæ passing from without inwards may with a high magnifying power be made out. Near the centre of the cell is the clear oval nucleus, bounded by a distinct outline, and containing generally one nucleolus. Globules of fat of varying size may, if the animal were killed during digestion, be seen within the cell; they are recognised by their strong refractive power.

Of the groups, some may occur in which the cells are seen from above: the collective bases will then present an appearance of polygonal areas intersected by lines of intercellular substance; in other groups, where the cells are seen laterally, their arrangement with regard to one another will be observed.

Some cells will probably be seen which have acquired a peculiar chalice-like shape, owing to the part of the cell near the free surface having become swollen out with mucus, often to the extent of bursting away and destroying the free border. These are the so-called 'goblet cells.'

The other two portions of tissue will not be ready for examination for two or three days. A small piece or scraping of the tissue in the chromic solution is teased in a drop of distilled water and covered, with a small hair under the cover-glass; and the small fragments are further broken up by tapping the cover-glass. Staining solution (carmalum or very dilute hæmalum) is now allowed to diffuse under the cover. This stains the nuclei and to some extent the protoplasm of the cells. When the staining is considered sufficient a drop of dilute glycerine is placed at the same edge of the cover-glass to which the stain was applied, and this also is allowed to diffuse under the cover, which may subsequently be cemented. Preparations are made from the osmic preparation by gently scraping the epithelial surface and mounting the product in dilute glycerine. The separation of the cells from one another is effected by tapping the cover-glass.

The cells will be found to exhibit the same general characters as regards form and appearance, but they are stained of a dark grey colour, and in consequence appear for the most part very distinct. Any fatty particles which they may contain are coloured intensely black. To preserve this preparation all that is further necessary is to apply a little fixing cement of some sort around the edges of the cover-glass.

Study of the finer structure of cells and nuclei. Karyokinetic figures.—The study of the structure of cells and nuclei, both at rest and during division, is in the first instance best made in the epithelial tissues of tadpoles of the salamander (*Salamander maculata*) or in the tail of the newt (*Triton cristatus*). To obtain salamander tadpoles the female salamanders are procured in January or February. They are viviparous, and at this season of the year are generally full of embryo tadpoles. The latter are placed in Flemming's or Hermann's solution for two days, and are then washed for several hours in running water, after which they may be preserved in a mixture of equal parts of glycerine, alcohol, and water. They may also be fixed and hardened in corrosive sublimate solution or in picric acid (see pp. 17, 18). The best parts to take from the tadpoles are fragments of epidermis from the end of the tail and pieces of the gills, or of the parietal peritoneum.

The following method is recommended by Flemming. Thin shreds or sections of the fixed and hardened tissue are stained for two days in a mixture of equal parts of alcoholic solution of saffranin (saturated) and aniline-water, then washed with distilled water, and differentiated with absolute alcohol, which may be acidulated with 1 part hydrochloric acid per 1,000; again washed in distilled water, and then placed in gentian violet (saturated watery solution) for one to three hours; after which they are again thoroughly washed with distilled water. They are next rinsed with alcohol until nearly all the stain is washed out, except from the nuclei, and the pieces are finally transferred to oil of bergamot, and mounted at once in xylol balsam. After removal from the gentian violet and washing with water they may be first differentiated in a saturated aqueous solution of orange G, and then passed through two changes of absolute alcohol into bergamot oil.

A thick piece of tissue such as the newt's tail must be cut into sections after hardening and embedding in paraffin and the sections fixed on a slide (p. 37); they may then be stained as above. Or the hardened tissue may be stained in bulk with aqueous solution of saffranin for two days, and after washing with water may be placed in gum and cut into sections by the freezing method. The sections are placed in water, transferred one by one to alcohol, and rinsed in this until differentiated. They are then transferred through oil of bergamot to xylol balsam as before. In these preparations the nuclei, especially those which are undergoing karyokinesis, are intensely stained; the nuclear spindles may be made out in some cells, and in some the attraction particles or centrosomes may also be detected, especially if Henneguy's method have been employed (see p. 21). Simply staining with very dilute hæmalum gives fairly good results for the chromatic nuclear structures.

After fixation in Hermann's fluid, followed by alcohol, and this by crude wood vinegar (twelve to eighteen hours), the chromatin elements may be sufficiently conspicuous without further stain. This method is recommended by F. Hermann for the study of karyokinetic figures in the testicles of salamanders killed about the end of July. Thin sections of the organs should be made by the paraffin method.

Fibres in epithelium cells.—The fibrous structure which certain epithelial cells exhibit may be seen and studied in the following way (Nuël):—Kill a pigeon and with a finely-drawn glass pipette immediately inject a drop or two of 1 per cent. osmic acid solution into the anterior chamber of the eye. Then cut out the cornea and mount it in osmic acid with the posterior surface uppermost. On this surface there is a layer of flat epithelium cells (endothelium of Descemet's membrane), and on examining these cells with a good microscope they are seen to be pervaded with fibrils which traverse the cells and also pass across the intercellular spaces from cell to cell. To preserve the preparation it may be washed after being in osmic acid for an hour, and mounted in dilute glycerine.

Intercellular substance. Silver method.—To show the intercellular substance of an epithelium (or endothelium) the fresh tissue must be taken, and after being rinsed with distilled water placed for a few minutes in 1 per cent. nitrate of silver solution. On removal from this it is again rinsed with distilled water and placed in tap water in the light. In a few minutes the intercellular substance becomes stained brown or black, and the preparation can be mounted in glycerine, or it can be passed through alcohol and clove-oil into xylol balsam.

Ciliated epithelium.—For the present we may conclude the study of epithelium with the description of the modes of viewing ciliated epithelium, and of studying the action of various reagents upon the ciliary motion. The other more specialised forms of epithelium, which are found in glandular organs and elsewhere, will be seen and studied when the several organs and parts in which they occur are prepared.

Ciliated epithelium in its living state may be readily obtained from the mouth and gullet of the recently killed frog. A drop of aqueous humour should first be collected by passing a capillary glass tube through the cornea into the anterior chamber of the eye; the drop is placed upon a slide, and then, the frog's mouth being held open by an assistant, the roof is gently scraped with the point of a clean scalpel, so as to remove the adherent mucus. A little of the epithelium will be brought away with this, and on placing it in the fluid and covering the preparation (taking the precaution of previously placing a hair in the drop), the cells may be sought for with a high power. For the most part they will be collected into little groups of three or more, the cilia being in active movement and producing currents in the liquid, so that free particles, blood-corpuscles, and other small objects are moved along in it. But if the group is small, or especially if entirely isolated cells are seen, it will generally be found that the cilia act upon the pieces to which they are attached like little paddles, moving them about in the fluid. The cells, it may be observed, are either shortly columnar or are spheroidal;

the nucleus is seldom distinct, because concealed by the granular nature of the cell-protoplasm. The cilia themselves can best be seen when they are moving languidly or when their motion has altogether stopped ; they are very fine, and spring, a number together, from the free surface of each cell.

Cilia of mussel.—But by far the most convenient object for the study of ciliary motion is to be found in the gill of the common seawater mussel (*Mytilus edulis*). Here the cilia are very large, and their motion will go on unimpaired for many hours. Hence they are particularly well suited for the

FIG. 47

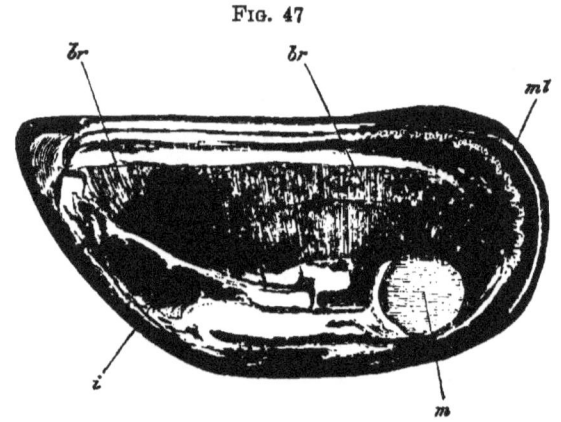

Valve of mussel, showing, *br*, *br*, the expanded gills or branchiæ, which, owing to the little bars of which they are composed, present a striated aspect

ml, mantle ; *m*, cut adductor muscle ; *i*, mass of viscera ; the dark projection just above is the foot

observation of the action of most of the reagents which affect ciliary movement.

One or more mussels may readily be procured at any fishmonger's ; those only should be chosen which remain tightly closed, for those with open valves are in most cases already dead. The valves may then be forcibly separated by means of a knife, when the gills (fig. 47 *br*.) will come into view, as flattened expansions of a yellowish colour, covering a considerable part of the shell, inside its lining membrane *ml*. By observing carefully it may be noticed that they have a striated aspect, the markings passing transversely to their length, and by taking up a small portion with a forceps it will further be seen that this striation is due to the fact that the gill is made up of a number of little bars which are distinct from one another for the greater part of their length. Take now a small piece of the gill, including three or four of the bars, and placing it upon a slide in a drop of the seawater which the shell always contains, separate the bars one from another by means of needles; the preparation may then be covered and observed.

Each of the bars in question will be seen to be fringed with large cilia, which are set at an appreciable distance apart along nearly its whole length, but at the free extremity of the bar are much more densely arranged. Those in this situation resemble in appearance the cilia of the frog's mouth, with the exception that they are very much longer; and like them they appear to spring a large number from each cell, whereas the others are stiff-looking and obviously thicker, and are connected at their base each to a single, comparatively small epithelium cell. In spite of these differences of appearance and attachment the two kinds seem to be essentially alike in nature, for the mode in which they move is similar and they are similarly affected by reagents.

It will be found that after a preparation such as that just described has been made for several hours, the movement will have become somewhat languid, and then the manner in which the individual cilia move can be more clearly made out. The preparation can be used also for the study of those agents which tend to revive and stimulate ciliary motion, and it will be seen that it is precisely the agents which most accelerate the amœboid movements of the white blood-corpuscles that have the most marked effect upon the cilia also.

Action of warmth upon ciliary motion.—The same mode of applying heat to a preparation of cilia is to be used as was employed for observing the effect of warmth upon the blood (p. 53). It is well, after enclosing the preparation with oil in the manner there detailed, to put it aside for some time, when it will probably be found, as just stated, that the movement of the cilia is languid or altogether arrested. On now gradually warming the preparation the motion becomes revived, and as the heat is raised becomes, *pari passu*, gradually faster, until a point is reached at which the cells are injured by the high temperature, when the movement slows and is arrested, and is not again resumed. But if the experiment be stopped short of this point and the source of heat removed, it will be seen that, conversely, as the temperature of the stage falls the rate of movement also diminishes, until, when again quite cold, the cilia may again almost stop, although they can be made to resume their active motion on again applying warmth.

Action of alkalies.—A very weak solution of caustic potash in salt solution or in seawater, similar to that which was used in investigating the effect of weak alkalies upon the blood, may be applied to a preparation of cilia which have become somewhat languid, in exactly the same manner as in the case of the blood—by allowing a little of the fluid to pass in at the edge of the cover-glass and diffuse itself with the seawater, so as to come gradually in contact with the slowly-moving cilia. The action is immediate, the cilia revive and vigorously lash the liquid into which they project, but the effort is soon exhausted, for the alkaline liquid penetrating the cells destroys their vitality, and the motion of their cilia stops beyond recovery. The stimulant action is not, however, peculiar to weak alkalies, for it is exhibited also by acids and by many other substances which, applied in stronger form, would instantly destroy the tissue, but when much diluted tend to revive and for a time maintain accelerated the ciliary motion.

Action of carbonic acid gas.—This reagent is to be applied to a preparation of cilia in the gas-chamber in the manner directed for the investigation of its action upon the newt's

ACTION OF CILIA

blood-corpuscles (p. 76). Everything being ready, choose a part of the preparation where the cilia are not acting very vigorously, and whilst still watching allow the gas to pass over the preparation. Its immediate action is seen to be that of a weak acid—that is to say, the rate of movement, if not already at its fastest, becomes accelerated—but as soon as the oxygen of the air in the chamber is entirely displaced by the continued stream of carbonic acid the motion ceases altogether. As soon as this result is obtained cut off the stream of CO_2

FIG. 48

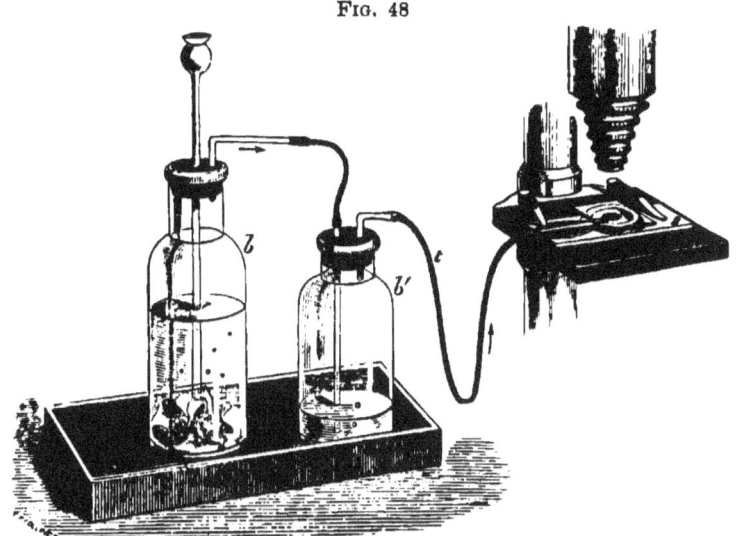

Apparatus for passing carbonic acid gas over a preparation under the microscope

b, bottle containing marble and hydrochloric acid ; b', wash-bottle ; t, india-rubber tube conducting the gas to the stage, s

and reverse the experiment by blowing air in at the side-tube, and thus displacing the carbonic acid from the chamber. The motion will almost instantly recommence. This shows that it was the absence of oxygen and not the presence of CO_2 which produced the stoppage of movement ; for there is, of course, an appreciable quantity of carbonic acid in the air which is thus blown from the mouth into the chamber.

Chloroform.—The gas-chamber is again used for this

reagent, the apparatus being arranged in the way previously recommended for blood (p. 61). Choosing a part of the preparation where the ciliary motion is vigorous, gently blow a stream of the mixture of air and chloroform vapour from the bottle into the moist chamber. The cilia become gradually slower and eventually stop. Now slip the indiarubber tube off the bottle and gently blow air through the chamber, to displace the chloroform vapour. The cilia will slowly revive on the readmission of air, and will soon be found to work as vigorously as ever. Like that with carbonic acid, this experiment can be repeated a number of times with a like result, if the chloroform vapour is not allowed to remain too long in contact with the preparation. Other vapours can be tested in the same way.

Isolation of ciliated cells.—To study the characters of the individual cells, a portion of membrane which is covered by ciliated epithelium is macerated in some fluid which softens and dissolves the intercellular substance, whilst preserving the cells themselves. The best for this purpose is a 2 per cent. solution of yellow chromate of potash. Chromic acid (1 in 2,000) or Flemming's solution, diluted 100 times with salt solution, or $\frac{1}{3}$ alcohol may also be employed. A large quantity of the fluid must be used, and the tissue—a piece of the trachea of a rabbit or other mammal, of the œsophagus of the frog or of the intestinal canal of a mollusc—is left in it for about forty-eight hours. With the point of a scalpel a little of the epithelium is then gently scraped from the inner surface, and being placed in a drop of distilled water on a slide, is broken up with needles as finely as possible. A small piece of hair is placed in the drop, to prevent the delicate cells from being crushed by the weight of the cover-glass. This is now superadded and the preparation carefully examined with a high power. If necessary the cover-glass may be tapped to separate the cells.

Numerous completely isolated cells are seen floating in the liquid, and these preserve for the most part their natural form and retain their cilia, although the latter are, of course, no longer in motion. The bright border through which the cilia appear to pass, the faintly striated cell-substance, the nucleus with bright nucleolus, and the truncated and often irregular fixed extremity of the cell are, in most, well exhibited. Besides these single cells others are present which are still united in groups or patches, in which, when viewed from the surface, the bases of the cells have a mosaic appearance. Moreover, mucus or 'goblet' cells may here and there be met with, and other cells with mucigen granules forming within them.

These are destitute of cilia; they lie, in the natural state, between the ciliated cells, and they may occasionally be seen *in situ* in some of the small groups of cells which have remained attached to one another.

If it be wished to permanently preserve such a preparation as that now under description, it is necessary first to stain the cells somewhat and then to substitute glycerine for the staining fluid. Either carmalum or dilute hæmalum or 1 per cent. osmic acid solution may be used for staining the cells. The two former mainly colour the nuclei very intensely, the last gives a uniform grey tint to the cells. The colouring fluid is applied in the following manner :—A drop is brought in contact with one edge of the cover-glass. When the staining fluid has diffused into the water in which the preparation was made, the preparation is left until the cells appear sufficiently coloured, a little more fluid being occasionally added if there seems any danger of the specimen becoming dry. With a solution of hæmalum, even though very dilute, a few minutes suffice ; the osmic acid solution should be allowed to remain an hour in contact with the cells. The staining fluid is replaced by a small drop of glycerine and water, which is applied at one border of the cover-glass and gradually takes the place of the water as this evaporates at the edges. The cover-glass may then be fixed.[1]

[1] For class purposes the method recommended on p 81 will be found valuable.

CHAPTER III

CONNECTIVE TISSUE

In the areolar tissue and in connective tissue generally there are several parts which present themselves for study ; and in order to observe each to the greatest advantage different modes of preparation are for the most part requisite.

The fibres of areolar tissue. Ranvier's demidesiccation method.—For the observation of the fibrous elements, simply, without special regard to their arrangement or relation to the other elements, all that is necessary is to place a small portion of areolar tissue, taken from any part, on the centre of a glass slide, just moistened with the breath, and with clean, sharp needles separate it as finely as possible into filamentous shreds. Then, without allowing the preparation to become dry, take a cover-glass on which a drop of salt solution has previously been placed and invert this over the tissue. The object of using but very little fluid to prepare the tissue in is to prevent the filaments from running together and becoming entangled when released from the needles.

In a preparation so made nothing is as a rule apparent save the wavy bundles of connective tissue fibrils, these when much developed obscuring the elastic fibres and corpuscles of the tissue. But if a second preparation be made in precisely the same way, except that, in place of salt solution simply, salt solution containing 1 part of acetic acid in 200 is placed upon the cover-glass, and if then the object is immediately examined with a high power, it is seen that the fibrils which compose the bundles gradually become indistinct.

whilst the bundles are soon much swollen, except, it may be, at intervals here and there. At the same time certain other fibres, almost equally fine but more sharply defined than the white fibrils, and always running singly, never in bundles, come into view. These are the elastic fibres. If one of them be followed for a short distance it will probably be seen that it sooner or later gives off a branch which unites it with a neighbouring fibre, whereas the white fibrils never show any disposition to branch or unite with one another, but those in each bundle maintain from end to end a perfectly parallel course. The elastic nature of the filaments which are brought into view by acetic acid is shown, in such a preparation as we are describing, by the fact that wherever in the process of teasing the tissue they have become broken across, the fibres have, by the recoil from the stretching to which they were submitted before the rupture occurred, been thrown into bold curves, especially marked near the broken extremities, which are often recurved. That this curved or coiled appearance of the elastic fibres, although highly characteristic, and always observable when the tissue is thus prepared, is, however, not a natural one, is shown by the fact that, as will immediately be described, when precautions are taken to preserve as much as possible the normal arrangement of the tissue-elements the elastic fibres are seen to pursue a rectilinear course.

The cells of areolar tissue.—To demonstrate the cells or connective tissue corpuscles the preparation is made more methodically. A film as thin as possible must be obtained for observation, so as to avoid the necessity of tearing the tissue. Such films are naturally present in the areolar tissue of most parts, and may be seen when the organs which it connects are gently drawn asunder from one another, as, for instance, when the skin is raised and reflected from the subjacent fasciæ and muscles. The most convenient source of such a delicate film is to be found in the exquisitely thin and transparent tissue which invests and lies between the muscles of the fore-limb of the rabbit and guinea-pig. The tissue in this situation, especially if taken from a young animal, is devoid of fat and not so completely overridden by the bundles of white fibrils but that the elastic fibres and the connective tissue corpuscles can be made out even without the addition of reagents. The mode of preparation is as follows :—

The animal having been killed by bleeding, the skin is snipped through around the upper part of the fore-limb and is then forcibly reflected from the limb. In this operation care must be taken to avoid besprinkling the subjacent parts with the cut hairs of the animal. A piece of the tissue over or between the muscles is then seized with forceps and snipped off with sharp, fine scissors. The snipped-off tissue shrinks immediately around the end of the forceps and appears very unsuited for microscopical examination. But place it on a clean slide, without the addition of any fluid, and with a pair of mounted needles endeavour, by drawing out first this corner and then that, to again reduce the gelatinous-looking piece to the condition of a thin film, and it will be found that this can be effected without much difficulty, for when not floated up by fluid the thin edges of the film tend to stick to the glass, and cease to shrink away from the position to which they are drawn by the needles. At the same time, whilst it is important not to *add* fluid to that which naturally moistens the piece of tissue, it is equally important never, during the whole process of stretching, to let the film become actually desiccated, for this would altogether ruin the tissue for microscopic purposes. The best way to prevent such a result is to breathe now and then on the object whilst it is being prepared ; by so doing needless haste will be avoided and more time and pains can be taken for the complete display of the film. This being effected, a cover-glass (which should have been previously cleansed and placed in readiness, with a drop of salt solution upon it) is taken, and quickly superposed over the film of tissue, which is thus prevented from shrinking up again into a shapeless mass. The fibres, both the white (in wavy bundles of various sizes) and the elastić, and the corpuscles may now be carefully observed, at first with the usual high power and afterwards with the highest obtainable, and some of the corpuscles should be sketched. Moreover, search may be made for lymph-corpuscles, a very few of which are generally to be found in the connective tissue; they are readily distinguished from the fixed corpuscles of the tissue by their smaller size—small, obscure, and generally multifid nuclei—and especially their amœboid movements, of which it is probable no trace will be apparent even to the most assiduous observation in the connective tissue cells proper.

Although both corpuscles and elastic fibres may be seen in a preparation of this kind made with an indifferent fluid, they are better seen if the white fibres are acted upon by acetic acid, and still better if this action is combined with that of hæmatoxylin, so that the corpuscles are brought more prominently into view. Moreover, the preparation admits of being permanently preserved in glycerine after such a method of treatment. Up to a certain stage the procedure is the same as that above described, but instead of placing salt solution upon the cover-glass before inverting it over the film, a solution of acetic acid (1 per cent.) is used, which tends to swell up the white fibres and thus brings the cells and elastic fibres clearly into view. The cells can then be stained by running hæmalum solution under the cover-glass. Very good results may also be got by applying carmalum (not hæmalum) directly to a film, without the previous action of acetic acid.

The connective tissue corpuscles can probably be made out at once in the thinner parts of the preparation, with their clear oval nuclei and the flattened irregular area occupied by their cell-substance. In a few minutes their nuclei will be tinted by the stain, and will then show up much more prominently; but to get the cell-bodies sufficiently coloured it will be necessary to leave the staining solution half an hour or more in contact with the preparation. Meanwhile, to obviate the effects of evaporation. a considerable drop of the colouring fluid should be placed on either side in contact with the edges of the cover-glass. The excess of fluid, moreover, has a tendency to raise the latter slightly from the film of tissue, and in this way a more ready access of fresh colouring fluid is permitted. When it is found on examination that the corpuscles are properly stained, the solution may be drawn off by a slip of filter paper applied to the edge of the cover-glass on one side, whilst to the other water is applied and frequently renewed, the excess of stain being in this manner rinsed away. A drop of glycerine and water coloured by magenta is now drawn under the cover-glass; the magenta serves to stain the elastic fibres, and the glycerine permanently preserves the preparation; the cover-glass may then be at once cemented. Picrocarmine may also be employed as a stain, but, as it only acts slowly, it is necessary to keep the preparation in a moist chamber for some hours to prevent the evaporation of the water.

In specimens treated with acid there may be observed a constricting ring at intervals along the course of some of the connective tissue bundles, an appearance which has long been familiar to histologists. The nature of this constricting ring is not yet clearly determined, some supposing it, from its resistance to the action of acids, to be of the nature of elastic tissue, others that the appearance is caused by the process of a corpuscle enwrapping the bundle. As shown by Ranvier, the constricting ring is, like the cells, tinted red by picrocarmine, whilst the elastic fibres are coloured yellow.

Preparation of denser areolar tissue.—In most of the larger animals (*e.g.* the dog) the connective tissue is, in the adult, so densely pervaded by bundles of white fibres as to render it impossible to obtain a film delicate enough for easy observation without tearing the texture with needles, and thereby distorting the cellular elements. Or it may be desired, even in those animals in which such delicate tissue as that the preparation of which has just been described is found, to obtain a specimen from a 'part where the connective tissue is not naturally extended in so advantageous a manner for preparation and observation. In these cases the following method may be employed with advantage :—

In a recently killed dog a flap of skin is dissected back, and a hypodermic syringe (fig. 49), provided with a fine cannula, having been previously filled with salt solution, the point of the cannula is inserted underneath the layer of connective tissue which is most superficial on the reflected portion of skin, and a little of the fluid is forced out. This does not immediately diffuse itself uniformly through the loose areolar tissue, but remains for a short while at the same place, forming a little bulla of liquid bounded and covered in by a film of tissue, the thickness of which depends upon

Fig. 49

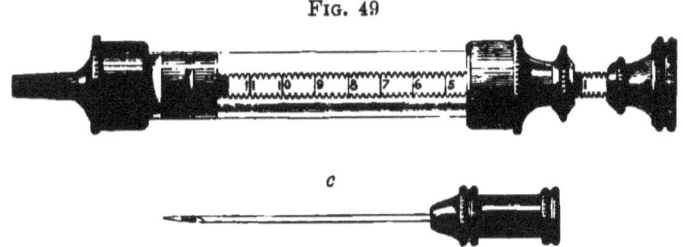

Syringe for interstitial injections; *c*, the cannula. The syringe is glass with ebonite fittings; the cannula is a fine perforated steel needle set in ebonite. Natural size.

the depth to which the cannula was inserted. If it does not appear thin enough a second attempt should be made at another spot. Then, before the bulla has time to subside— that is to say, before the fluid has time to diffuse itself through the meshes of the tissue—snip off the whole projection with a single cut of a pair of scissors, which for this purpose should be particularly sharp and clean, and transfer the snipped-off portion to a clean slide. Here it may either be at once covered in salt solution and examined without reagents, or may be treated with acetic acid, hæmatoxylin, &c., as in the mode of preparation just described.

A modification of the above method consists in injecting into the tissue a solution of gelatine instead of salt solution.

The gelatine solution is made by taking some clear French gelatine, allowing it to soak for an hour or two in water, and then, after pouring off all the excess of that fluid, placing the soaked gelatine in a beaker over a water-bath until it is entirely melted in the water which it has imbibed. The syringe is then warmed by immersing it for a minute or two in warm water, and is filled with the gelatine solution, and a little of this is injected into the subcutaneous connective tissue, so as to produce a bulla like that made by the salt solution. In cold weather the gelatine will set almost immediately; in warm weather the process may be accelerated by surrounding the bulla with small pieces of ice.

When the gelatine is quite firm, sections of it are made with a razor. As they are cut they are placed in salt solution.

Before mounting them it is well to stain the specimens, and this may advantageously be done by aqueous solution of magenta. This colours the elastic fibres strongly, the corpuscles distinctly, and the bundles of white fibres slightly, while the gelatine which was injected, and of course occupies all the interstices, is hardly stained at all. The time of immersing the sections varies of course with the strength of the fluid, but this should not be too highly coloured, and it can be then seen without much difficulty when the sections are sufficiently stained. They are subsequently placed in water for a minute or two to remove the excess of magenta prior to transferring them to a slide. A drop of glycerine is now added and the cover-glass laid on, after which the slide is gently warmed over a small flame or otherwise until the gelatine in the sections just melts, so as to allow the cover-glass, which was probably tilted up somewhat owing to the thickness of the sections, to settle down. The specimen may then be examined, and if satisfactory may be preserved, the preparation being completed by fixing the edges of the cover-glass with gold size.

By the modes of preparing connective tissue already described, most of which we owe to Ranvier, the bundles of white fibres, the elastic fibres, and the corpuscles are brought under observation, and it would seem at first sight that these of themselves entirely constitute the tissue. But in considering the nature of the films obtained—that they are, namely, continuous over a greater or less area—it is clear that the presence of fibres and cells is not alone sufficient to account for the laminæ which are spread upon the slide. And, indeed, by closely observing the preparation it will be apparent that there is pellucid substance uniting everything together, through which the fibres run, and in which the corpuscles lie embedded. There is, it is true, a difficulty in making this out in most parts, in consequence of its extreme clearness, and the fact that its refractive index is little different from that of the watery fluid the tissue is examined in; moreover, in the logwood preparations the intermediate substance remains entirely unstained. Nevertheless, towards the edge of the preparation, where a comparison can the better be made with the surrounding fluid, the fact that such a clear intermediate substance does really exist will be sufficiently evident; and the more so if the cover-glass be slightly moved, or one edge be gently pressed down with a needle. But we possess in the silver method of Recklinghausen a ready means of demonstrating its existence in an obvious manner; for the ground-substance (or intercellular substance) of the connective tisue, and, indeed, of almost every other tissue in the body, possesses the distinctive property of reducing the salts of silver under the action of light, so that the metal is deposited in it. The effect of this deposition is to stain the ground-substance of a colour varying with the intensity of the light employed and with other conditions from a light brown to a brownish black. The fibrous elements participate for the most part in this staining, and are frequently, especially when the preparation has been, as is usual with silver preparations, mounted in glycerine, indistinguishable from the ground-substance through which they course, and which also unites the white fibres into the bundles which they form. The cellular elements, on the contrary, remain absolutely unstained, and, moreover, after the action of the silver salt are no longer affected by those staining fluids which otherwise have a particular affinity for them; it is therefore no longer possible to bring them into view. Wherever, then, a cell is situated, there appears after the reduction of the silver nothing but a white patch upon or in the brown ground; and if, as is not unfrequently the case, several flattened cells may have occurred together with their edges in juxtaposition, the group appears as a larger white patch intersected by dark lines, these representing a small amount of intercellular substance between the individual cells. The appearance is similar to what is observed in an epithelial tissue after the silver treatment, for in this the intercellular substance is always very small in amount. Such an arrangement of connective tissue cells is on this account designated 'epithelioid.' The white patches, then, in the silvered preparation of connective tissue represent either depressions on the surface of, or actual cavities within, the matrix or ground-substance, containing cells, which themselves are not visible, so that the white patches are termed the *cell-spaces* or (recalling the analogous case of bone) the *lacunæ* of the connective tissue.

It is the more appropriate to give them a special designation because, as may be made out by a careful comparison of specimens of connective tissue from the same part, some prepared with chloride of gold, to show the cells, others with silver, to show the cell-spaces, the cell-spaces are in many cases distinctly larger than the cells; they are not necessarily therefore, as has sometimes been supposed, and as is no doubt the case with the clear part of a silvered epithelium, merely the cells left white. The difference in the relative size of the cell-spaces and the contained cells obtains no doubt more frequently, or at least can be more readily made evident, in the firmer varieties of connective tissue, where the ground-substance is everywhere pervaded with fibrous bundles, and has in consequence lost its soft and yielding nature, which otherwise permits it to adapt itself more readily to the shape of the cells.

Thus much having been said in order to explain the appearances produced by the silver method of treatment, the best mode of applying it to ordinary connective tissue, such, for instance, as the subcutaneous, may now be described.

Application of silver method to areolar tissue : cell-spaces and intercellular substance.—The skin of a recently killed rabbit or guinea-pig [1] having been stripped off one of the limbs,

---

[1] These animals are selected because there is likely to be less fat in the subcutaneous tissue than in that of the cat or dog or other animals commonly used in the laboratory. It is important to remember that any tissue which is to be submitted to the silver method must be fresh and unacted upon previously by any reagent whatever; moreover any blood on the part must be rinsed away with distilled water.

this is disarticulated at the proximal joint, and is rinsed for a second or two in a beaker of distilled water, in order to wash away any blood or lymph which might happen to be on the surface, and which would cause a granular precipitate with the nitrate of silver solution. The latter, a solution of 1 part of the salt to 100 of distilled water, is then either poured over the surface or dropped on it from a pipette. After two or three minutes the silver solution is quickly washed off by a stream of distilled water, and the limb is then at once placed in a beaker of spirit, and exposed to direct sunlight, or, failing this, to bright diffused daylight. In a few minutes in the sunlight, and after a longer time in diffused daylight, the silvered surface will have acquired a uniform brownish tinge to the naked eye. When the colour is strongly marked it is as well to remove the beaker from the light, lest the preparation become too darkly stained. The limb should be allowed to remain in spirit during twenty-four hours ; at the expiration of this time it is placed in a dish, and, by the aid of fine forceps and scissors, pieces of the superficial stained layer are dissected off under spirit. In doing this care must be taken not to drag at all upon the membrane which is thus removed, so as to throw it into creases. Or thin slices may be cut with a razor from the surface browned by the silver. The pieces are then transferred to oil of cloves on a clean glass slide ; and most of the superfluous oil of cloves having been wiped away, a cover-glass, on which a drop of xylol balsam has been placed, is inverted over the preparation. Before putting on the cover-glass it is well to examine the object under a low power, in order to make sure of the absence of folds and creases or specks of dust upon it : if any such be seen they must be carefully removed with a needle. Indeed, it may be recommended as a golden rule in making histological preparations never to put the cover-glass in its place until a glance at the object under a low power of the microscope has certified the absence of any marked imperfection : if this be attended to the time will often be saved which would otherwise be spent in mounting worthless specimens.

A simpler method of showing the ground-substance and cell-spaces of the subcutaneous tissue consists in making a film by the method of demidesiccation as detailed on p. 98, and placing a drop of 1 per cent. nitrate of silver solution upon the middle of the film. After a minute or two the silver nitrate is rinsed off with distilled water and the preparation is placed for a few minutes in sunlight. The water may then be removed by alcohol, this replaced by clove-oil and the preparation finally mounted in xylol balsam.

The connective tissue which covers the tendons of the superficial flexor digitorum of the ox's foot, as they run through sheaths formed by the tendons of the deep flexor, is more easily prepared by the silver method than the looser kinds such as the subcutaneous tissue. A piece of such a tendon is taken, rinsed in distilled water to remove the synovial fluid which covers it, and treated with silver in the way described on the preceding page; and is then, after washing, placed in the light in spirit. It soon becomes brown, when it may be removed from the light; and after remaining twenty-four hours in the spirit it is easy, with a sharp knife or razor wetted with spirit, to obtain a thin surface section. This, after being immersed for a minute or two in clove-oil to get rid of the spirit, is mounted in xylol balsam with the browned surface uppermost. It should present, if successful, an extremely characteristic and beautiful image of white branched cell-spaces, single or in groups, upon a brown ground.

In both the preparations last described it is possible to show the nuclei of the corpuscles which lie in the cell-spaces by subsequent staining with logwood. Sometimes they are visible even without this treatment.

Elastic network of areolar tissue.—The proportion of elastic fibres varies considerably, according to the part from which the connective tissue under examination is taken. Some serous membranes contain a large number of elastic fibres ; and since they are readily spread out in their natural condition the network which these fibres form by their branchings and conjunctions is easily made evident. One of the best objects for this purpose is to be found in the rabbit's mesocolon. A piece of this, moistened with a little salt solution, may be spread out as flat as possible upon a slide, and a drop of dilute magenta solution having been placed on a cover-glass, this is inverted over the tissue. The elastic network becomes stained by the dye, whilst the white fibres remain almost unstained. The preparation may be made permanent by putting, as in former preparations, a drop of dilute glycerine at the edge of the cover-glass, and after this has had time to diffuse itself cementing the edges with gold size.

Elastic tissue.—The elastic ligaments are connective tissue structures in which the elastic elements of the tissue greatly preponderate. There is always a quantity of ordinary areolar tissue amongst the fibres, but not sufficient to obscure them, especially as they are generally of larger size as well as in greater number than elsewhere. It is sufficient, in order to see them, simply to teaze out a portion of the ligamenta subflava of the vertebræ or of the ligamentum nuchæ of the ox in water or salt solution. If it be desired to keep the preparation it can be mounted in glycerine or glycerine-jelly.

Section of elastic fibres.—To observe the shape of the fibres a transverse section may be made of a piece of ligamentum nuchæ of the ox, in which the fibres are extremely large.

In order to obtain the requisite firmness for cutting, place a small piece of the ligament in a quantity of 2 per cent. solution of bichromate of potash for fourteen days ; then in water for two or three hours ; then transfer to gum. When soaked with this, sections across the direction of the fibres may be made with the freezing microtome. The sections are to be placed in water to remove the gum, and mounted in glycerine, either with or without staining with magenta or hæmatoxylin.

Fibrous or tendinous tissue.—This may be examined by separating a small shred from a tendon or ligament, and teasing it out as finely as possible into its constituent bundles. The operation is conducted with the aid of needles by the method of demidesiccation, and it is first examined in salt solution, being afterwards treated with dilute acetic acid and hæmatoxylin. But it is a troublesome matter to make the separation fine enough without disturbing too much the arrangement of the cellular elements of the tissue. Fortunately we can obtain, from the tail of the mouse or rat, tendons which are, so to speak, naturally dissociated; for fine silk-like tendons run along the whole length of the tail, and can readily be drawn out, needing no further manipulation than is necessary to place them advantageously under the microscope. The following is the mode of procedure :—

In a recently killed mouse the tail is seized about half an inch from the tip between the thumb-nail and fore-finger of the right hand; and the delicate skin being nipped through and the vertebral column broken at this point by the pressure of the nail, it will be found quite easy, the base of the tail being fixed by the left fore-finger and thumb, to separate the end altogether and drag it away from the remainder of the tail. In doing so it will be found that the minute tendons which are attached near the tip, owing to their comparative toughness and strength, are not broken through at the spot in question, but are dragged out of the channels in which they run, and may in this way be obtained in the form of a bundle of exquisitely fine silky threads, which are to be immediately immersed in a glass dish of salt solution. Now cut away two or three of the fine threads with sharp clean scissors, and seizing them by one end with fine forceps, or leading them with a needle-point, float them on to a glass slide which is held immersed in the fluid, and is then carefully lifted out. After arranging the minute tendons as nearly straight as possible on the slide and blotting up most of the superfluous salt solution or allowing it to run off, place a short piece of hair parallel with them, to avert the pressure of the cover-glass, which is now placed over the middle of the threads in such a way that, since they are considerably longer than the width of the cover-glass, their ends project beyond on either side. The object of this is to permit them to be drawn straight should the superposition of the cover-glass have displaced them. These ends, moreover, since they are exposed to the air, soon dry and stick to the slide, so that subsequent treatment with reagents does not tend to displace the tendons, which are thus maintained in an extended condition. Examined thus in salt solution, little is visible beyond the slightly wavy, closely packed white fibrils, collected, as longitudinal streaks seen here and there indicate, into a few indistinct bundles. But allow a little dilute acetic acid (1 part of the glacial acid to 200 of salt solution) slowly to pass under the cover-glass, and a remarkable change becomes apparent. As the acid reaches the tendons they slowly swell up and become more transparent, the fibrils becoming indistinct; and now chains of small oblong faintly granular cells, each with a clear nucleus situated near one end of the cell, and often opposite that of a neighbouring cell, come into view. These are the *tendon-cells*, the corpuscles of the fibrous connective tissue; only the central thicker portion of each, which lies in the interstice between three or more tendon bundles, is seen at present; the thin lamellar prolongations, which extend between two tendon bundles, are too delicate to be made out without staining. In some of the chains a bright longitudinal line is to be seen on each cell; this appearance is merely produced by a lamellar prolongation of this sort which happens to extend vertically to the plane under observation.

After the action of the acid has been prolonged for some time, the cells gradually lose their distinctness, and eventually can with difficulty be made out, although the fibres are more swollen and indistinct than ever. But if a little carmalum or hæmalum is allowed to run under the cover-glass, first the nuclei, and then the bodies of the cells and their prolongations become coloured, whilst the fibres remain unstained, just as in the parallel case of the areolar tissue.

When the colouration is sufficiently deep, the staining fluid may be replaced by water, and this in the usual way by glycerine and water ; and finally, the edges of the cover-glass being cemented, the preparation can be permanently preserved. Examined with a high power, the tendon-cells now appear in the successive horizontal planes as quadrangular flattened bodies, thickest near the middle, and gradually shading off at either side, and marked with one or more dark lines running longitudinally, which are, in fact, the bright lines to which attention was previously drawn, and which have now become stained. To show that these are actual flattened extensions of the cell, and not mere markings, it is necessary to compare the appearances presented by a transverse section.

Another method of displaying the cells of tendon is that originally employed by Ranvier. One or two of the small tail tendons are placed on a slide, and their ends are fixed with paraffin, so as to keep them extended. A few drops of a 1 per cent. solution of picrocarminate of ammonia are then placed upon them and left for from half an hour to several hours, after which the picrocarmine is washed away with distilled water and the tendons are mounted in glycerine, acidulated with acetic acid.

Transverse section of tendon.—To obtain this, sections may be made across a piece of the tail of the mouse or rat. For observing the cells the tissue should be stained with gold chloride as follows :—A short piece of the tail deprived of its skin is immersed for five minutes in formic acid solution (1 to 4). It is then placed directly in gold chloride solution (1 per cent.) and left in this for about an hour. Then it is rinsed with water and replaced in 1 to 4 formic acid, in which it is left, in the dark, for twenty-four or forty-eight hours. It is now thoroughly washed with water, soaked in weak gum, frozen, and sections are cut across it with a microtome. The sections should be mounted in glycerine.

For most purposes, however, it is best to take a large tendon, for it is easier to get transverse sections of such a one, and in all essential points of structure it is quite similar to the minute tendons which, for the sake of convenience, we have just been employing. A piece, then, of any tendon, large enough to be grasped by the fingers, is placed in strong spirit for a day or two. This gives it a very hard, horny consistence, and it is easy, with a sharp knife or razor wetted with spirit, to get one or two thin sections from the end. These are placed in aqueous solution of magenta, or picrocarmine, or in carmalum until they are sufficiently stained. The excess of stain is then washed away by water; and finally a little diluted glycerine is placed upon a cover-glass, which is inverted over the preparation.

Another and perhaps a better method is to harden the tendon in 2 per cent. bichromate of potash for a fortnight or more, and after soaking in gum to cut sections of it by the freezing microtome. The sections may be stained with carmalum and mounted in xylol balsam after being passed through alcohol and clove-oil.

It will be seen that the tendon is divided into fasciculi by septa of areolar tissue, the corpuscles of which (seen edgeways) are brought into view, being stained by the logwood; it will further be observed, if the sectional area of one of the smaller fasciculi is attentively examined with a high power, that it again is divided (although incompletely) into several still smaller bundles by the branching processes of deeply coloured stellate bodies situate at the angle of junction between three or more such bundles, and extending a greater or less distance between the neighbouring bundles. These stellate bodies, with their processes, are the tendon-cells with their lamellar extensions, as seen in section; the smaller fasciculi, which are separated by areolar tissue, correspond with the whole of one of the caudal tendons of the mouse or rat.

Cell-spaces of tendon.—To show the cell-spaces in which the above-mentioned tendon-cells are contained :—

As before, break off the end and draw out some of the

tendons of a mouse's tail (that previously used will still yield a sufficient number). Then hold the tendons in a shallow dish of distilled water, and with a medium-sized camel-hair pencil brush them firmly from end to end six or eight times. Remove them now from the water, and immerse them in a large watch-glass of nitrate of silver solution (1 per cent.) for five minutes; then place in a glass vessel of distilled water and expose to sunlight. As soon as they are brown, pieces may be cut off, laid straight in glycerine upon a slide and covered; after which the cover-glass may be fixed in the usual way. Or they may be placed (stretched) in alcohol for a few minutes, then passed through oil of cloves, and finally mounted in xylol balsam.

Epithelioid covering of tendons.—The object of first brushing the tendons is to remove the layer of flattened cells which covers the surface of each, and which, if allowed to remain, prevents the silver solution from properly acting upon the deeper parts of the tissue. To show this layer, another set of tendons may be treated with the silver solution in a similar way, but with a minute's immersion and without previously brushing them, when it will probably be found that the superficial epithelioid stratum is alone apparent.

Adipose tissue.—The simplest way of showing the fat-vesicles is by teasing out a small portion of the tissue in a drop of salt solution; taking the precaution of putting a narrow slip of blotting-paper on either side to avert the pressure of the cover-glass.

But a far better way of demonstrating the structure of the tissue generally, consists in the employment of the method of interstitial injection with gelatine described on pp. 101, 102. The gelatine is injected into the interior of a fat-lump, and sections are made and treated in the way described for areolar tissue. By this method the fat-cells are somewhat separated from one another, and all their parts, as well as the intermediate tissue and blood-vessels, are much better displayed.

Fat may also be studied by the employment of the other methods recommended for areolar tissue, especially that given on p. 98.

Membrane of the fat-cell.—To show this distinctly Ranvier's method may be recommended. An interstitial injection is made with a weak solution of nitrate of silver (1 in 1,000), and a minute portion of the fat-lump thus rendered œdematous is removed with scissors, transferred to a slide and covered. Many of the fat-cells exhibit the envelope and nucleus separated by a distinct space from the fat-drop. The silver solution would appear to have penetrated by endosmosis and to have become collected between the fat-drop and its enclosing membrane.

Development of fat.—To complete the study of adipose tissue, the fat-cells should be observed in process of development. For this purpose a preparation of the subcutaneous tissue from a part where fat is being deposited may be obtained from the moderately advanced fœtus of any mammal. The new-born rat is especially to be recommended, since in its subcutaneous tissue there are generally to be found, not only cells which are in every stage of fat-deposition, but others in addition which exhibit the formation of blood-vessels and the simultaneous formation within the same cells of red blood-corpuscles. The mode of preparation is very simple, all that is necessary being to strip the skin from the back, snip off with scissors a little of the gelatinous-looking tissue from the borders of a tract where the fat is already partly deposited, and to place the portion so obtained in a drop of salt solution upon a slide, and cover it.

CHAPTER IV

CARTILAGE

Articular cartilage.—This is to be studied in sections made both parallel and vertical to the surface.

Study of living cartilage.—From an animal that has just been killed remove one of the limb-bones, with its articular ends, and with a clean, sharp scalpel or razor take a slice, as thin as possible, off the cartilage, and quickly, before it has time to become dry, transfer the piece to a drop of serum upon a slide, place a cover-glass over the preparation and examine with a high power. Turning the attention more particularly to the cartilage-cells, the arrangement of these in groups in the faintly granulous matrix will be noticed. Each cell is seen to be provided with a clear round nucleus, which in some specimens of articular cartilage is so large proportionately that it may be mistaken by an inexperienced observer for the whole cell. In reality, however, the cell-substance is represented by the clear material (showing, at most, a few highly refracting granules) which lies around the nucleus and entirely fills the cavity or space in the cartilaginous matrix in which the cell lies. But now replace the serum by distilled water, drawing the former away by means of a piece of blotting-paper placed at one edge of the cover-glass, and allowing a drop of water from a pipette to run under at the opposite edge (see fig. 12), and the picture soon changes. Examine the cells at the borders of the slice, for these are first reached by the water. It will be seen that the clear cell-substance begins to be separated from the matrix, and acquires a jagged outline, fluid collecting in the interspace which is now left. This clear fluid is, there can be little doubt, derived from the cartilage-cells themselves, and not from the water which has permeated the matrix. The most probable explanation of its appearance is that the cartilage-cells, by virtue of a certain amount of vital contractility which they retain, shrink on contact with the water, as they do on the application of many other reagents and on the passage of an electric shock, and that the clear fluid which collects around them is expressed from their protoplasm as it contracts. Whatever the explanation may be, the effect is this : that the now jagged, shrunken cell-body assumes, instead of the clear aspect which in the fresh condition it presented, a coarsely granular appearance, so much so indeed that the nucleus which was previously so apparent is now entirely obscured. Moreover, as already indicated, the cartilage-cell no longer fills the cell-space in which it lies. The cells always undergo this change after death, unless the tissue has been treated with some reagent which prevents its occurrence.

In preparing a specimen of cartilage with the object of permanently preserving it, our aim should be, as, indeed, with every tissue, to obtain it in a condition and form as nearly as possible approaching that which it had whilst living. There are numerous reagents which, in place of acting like water and causing contraction of the cartilage-cells, fix them in the form they present during life. Amongst these osmic acid may be mentioned first, as the most generally valuable reagent which we possess for this purpose, since it acts in like manner upon nearly all the tissues. Other and cheaper reagents serve the purpose equally well, however, for cartilage. One special value it certainly has ; namely, in showing that the little granules in the protoplasm of the cartilage-cell are many of them of a fatty nature, for they are blackened by the reagent. A 1 per cent. solution of alum, and a saturated solution of picric acid (both recommended by Ranvier), preserve the cells of cartilage admirably. But the best and most convenient reagent for the purpose is a weak solution of chromic acid (1 part to 500 of water).

Structure of articular cartilage.—The articular head of one of the long bones is removed from the recently killed animal, split into two down the middle with a strong knife or a chisel, and the halves placed in a large quantity of a solution of chromic acid of the strength above indicated, and allowed to remain in it a few days. The exact time is immaterial, but in specimens which are left rather longer in the liquid, the bone in the neighbourhood of the cartilage is softened, and the cartilage-cells are more coloured by the acid, and consequently more apparent. When it is desired to prepare the tissue for the microscope, one of the halves is taken from the fluid, washed in water for a minute or two, and then, the bone being held in the hand, one or two thin sections are taken from the surface of cleavage, and therefore vertically to the articular surface, including, if possible, a little of the adjacent bone. The sections are placed in water on a slide and covered, and are examined first with a low power to study the general arrangement of the cartilage cell-groups in the superficial, intermediate, and deep strata respectively, and subsequently with a high power to see the intimate structure of the cells. These should, as before indicated, present as nearly as possible the same appearance as during life, the only difference being that the tissue generally is less transparent, and slightly coloured, and that the cell-outlines are rather more strongly marked. These differences become less obvious when glycerine has been permitted to diffuse itself under the cover-glass, for the preservation of the specimen. Previously, however, to the addition of glycerine the preparations may be stained by hæmalum or carmalum, after well washing them with water.

Sections are next to be taken parallel to the articular surface and mounted in the same manner; but it must be borne in mind that it is only those sections which include parts of the cartilage near the natural or artificial surface which will be of value as respects the preservation of the tissue elements in their natural condition, at least in the case of the thick articular cartilages of large animals. For the preservative solution naturally takes some time to permeate the cartilaginous matrix, and before it has time to penetrate to the deeper parts the cells will have already shrunk away from the walls of the enclosing cavities and have become changed in the manner previously indicated. So that the deeper sections will exhibit merely the irregular, contracted, and highly refracting corpuscles lying loosely in their cell-spaces.

At the thinnest parts of all the sections cavities may be observed in the matrix which are devoid of cartilage-cells; these having dropped out in the process of preparation.

Cell-spaces of cartilage. Preparation by silver nitrate. The cavities or cell-spaces of cartilage may be readily shown by the same method as was employed to show the cell-spaces of connective tissue, viz., treatment with nitrate of silver and subsequent exposure to the light. For this purpose a fresh joint should be opened, and the articular end of one of the bones (preferably a *convex* one) removed with the saw or bone-forceps. The end thus removed is well rinsed in distilled water and then transferred to nitrate of silver solution (1 per cent.), in which it is allowed to remain three minutes. It is then again rinsed in water, and gently brushed with a camel-hair pencil to remove adhering silver precipitate; after which it is placed in a beaker of 50 p.c. spirit and exposed to sunlight. When thoroughly browned, sections are made from the surface with a razor wetted with spirit, and are placed in water (care being taken that they become completely immersed), after which they may be mounted in glycerine (or in xylol balsam after having been passed through alcohol and clove-oil).

Some of the sections should be taken from near the edge of the cartilage and mounted on a separate slide. These should show the branched cell-spaces, which present a transition to the much more ramified spaces of the connective tissue of the synovial membrane.

The cells themselves which occupy the spaces are not shown, for, as in silvered tissues generally, the cell-protoplasm remains absolutely unstained; indeed, it is at first sight difficult to believe that the rounded cavities which are seen really contain cartilage-cells. But their nuclei may be brought into view by staining with hæmatoxylin, subsequently to the matrix being coloured by the nitrate of silver. As before mentioned in speaking of connective tissue, it is difficult, if not impossible, at the same time to bring distinctly into view the outlines of the cell-spaces by the silver method, and the bodies of the cells themselves by some other method of staining.

Preparation of cartilage with gold chloride.—The cells of cartilage as well as most protoplasmic structures are well displayed when stained by the gold method. This is applied to cartilage in the following way. Thin sections are made from the articular end of a fresh bone and are placed in a few drops of a solution of chloride of gold (1 part in 200 of water). After half an hour they are transferred to a beaker containing a comparatively large amount of distilled water which has previously been slightly acidulated with acetic acid, so as just to be distinctly acid to the taste. The beaker is then covered with a glass plate and placed in a window in a warm place, and where it will be exposed for some hours a day to sunlight. Here the sections are left for two days, after which time they should have acquired a dark violet colour, and are ready to mount in glycerine. But before this is done they should first be examined with a low power in a drop of water to see whether there is any precipitated matter upon the surfaces of the sections. If this is the case it must be brushed away with a camel-hair pencil, the sections being held with fine forceps during the process. When finally mounted and examined with a high power, the cartilage-cells should retain precisely their natural form and appearance, except that they are stained of a violet tinge, whilst the matrix remains almost entirely colourless. In sections taken from the edge of the cartilage, the ramified transitional cells which occupy the corresponding cell-spaces shown by the silver method will be rendered apparent.

The ensiform cartilage of the newt is thin enough to be prepared by the gold method without the necessity of cutting

FIG. 50

Warming apparatus for maintaining portions of tissue exposed to the light at a constant, raised temperature. The lower part of the apparatus is filled with water, into which the gas regulator dips.

into sections. But it is well to strip off the perichondrium before the preparation is mounted in glycerine.

The colouration by the gold and silver methods appears to depend on the occurrence of a deposition of the metal in those parts of the tissue for which their salts have the greatest affinity: in the case of gold this is generally the protoplasm of the cells (and the nerve-fibres, where any exist) ; in the case of silver it is the ground-substance, or matrix, or intercellular substance ; so that the results of the two processes may, at least in the case of the cartilage-cells, be looked upon as standing to each other in the light of a positive and negative image, using the terms as they are employed in photography.

If it is the winter time, and especially if there is very little bright sunlight, it is of importance to keep the beaker of acidulated water in which the tissue is placed *warm*, as this materially facilitates the reduction of the gold. For this purpose it is well to have some arrangement by which the beaker, or several of them, if there be much of the same sort of work going on, can be kept at a temperature of about 30° or 40° Centigrade, and at the same time freely exposed to the light. Fig. 50 represents a convenient apparatus for this purpose. It is made of zinc or copper with glass sides and top, and may be fitted with a gas regulator to prevent the temperature from rising too high. When used for the present purpose the apparatus should of course stand in a good light—in a window if possible.

Costal cartilage.—The cartilages of the ribs and those of the trachea and larynx may be prepared and examined in a manner similar to that recommended for the study of articular cartilage. Sections, both parallel and vertical to the surface, should be made from the fresh tissue and from pieces which have been in chromic acid solution ; those from the latter being mounted and preserved in glycerine as before, with or without previous staining with hæmateïn.

Structure of cartilage matrix.—Sections are to be made of a piece of thyroid cartilage that has been preserved in spirit, and are to be stained with logwood. When sufficiently coloured, they are transferred to water and then mounted in glycerine. The logwood, besides staining the nuclei of the cells, gives the matrix also a deep purple colour. But this colouration of the matrix is not uniform, for some parts become stained much more deeply than others : those regions more immediately around the cells and cell-groups, and which therefore are, as commonly considered, the latest formed portions, having apparently more affinity for the colouring matter of logwood than the other and older parts. By this method the whole matrix appears marked out into what may be termed *cell-territories*, although series of definite rings can by no means be said to be very apparent around the cell-groups.

To show the fibrous structure of cartilage matrix, sections of fresh cartilage are macerated for some days or weeks in brine. They are then mounted, and pressure is made upon the cover-glass so as somewhat to crush the sections. A fibrous appearance may thereby be rendered apparent.

Transition to yellow fibro-cartilage.—If the arytenoid cartilage of the ox is sliced longitudinally, it exhibits to the naked eye in its lower part the opaline bluish appearance of hyaline cartilage, but in its upper part the faintly yellow aspect of elastic cartilage, the two parts being separated by a distinct line of demarcation. Take now a thin section from the cut surface, including a little of both parts, and mount, if fresh, in alum solution, or, if the cartilage has been previously hardened in spirit, at once in diluted glycerine, coloured by magenta. It is immediately seen that the cartilage above the line of demarcation merely differs from that below by the superaddition of a network of comparatively coarse branching (elastic) fibres. Just at the junction of the two parts the fibres are but few in number, and at their ends often imperfect—*i.e.* they may appear continued merely as rows of granules as if not yet fully formed. Further into the yellowish part they permeate the matrix very thickly, and give it almost a granular appearance, owing to many of them being seen in section. But immediately around each cell there is usually an area of the matrix entirely free from fibres, and this area is in appearance exactly like the matrix of hyaline cartilage. So that we may conclude that yellow or elastic cartilage is merely to be regarded as hyaline cartilage in the matrix of which a deposit of elastic substance has taken place in the form of branching fibres; and, indeed, the study of its development shows this to be the case.

Yellow fibro-cartilage.—A tissue may now be studied which is composed entirely of yellow cartilage, the human epiglottis or the external ear cartilage of the ox, for example. A section of epiglottis stained with magenta and mounted in dilute glycerine shows what at first looks like a granular matrix dotted with islands of hyaline substance, each with one or two cartilage-cells in the centre. The granular appearance of the matrix, when carefully observed, is seen to be due to an extremely close feltwork of fine elastic fibres; in fact, the structure is quite like that of the upper part of the ox's arytenoid, although, owing to their fineness and closeness in the epiglottis, the individual fibres cannot be followed for any distance.

White fibro-cartilage is best studied in sections of intervertebral disc or of some other ligament which is composed of this tissue. The tissue may have been previously hardened in spirit. Sections should be made both parallel with and across the bundles of tendinous fibres. They are to be stained with hæmalum and mounted in glycerine. It will be seen in the longitudinal sections that the cartilage-cells lie in chains between the tendon-bundles, occupying the place of the ordinary tendon-cells; from these they are chiefly characterised by their sharpness of outline and (in transverse sections) the absence of lamellar extensions, as well as their greater thickness; but transitions between the two are not unfrequent, especially near the tendon insertion.

CHAPTER V

BONE

Transverse and longitudinal sections of hard bone.—These are first cut as thin as possible with a fine saw, and are afterwards reduced in thickness and polished by grinding. The bone selected should be thoroughly macerated and bleached ; it should be absolutely free from grease. Sections may be made from different bones, flat and long, but for typical specimens of the compact tissue, transverse and longitudinal sections of one of the long bones of the limbs—the ulna, for instance—may be recommended, and a vertical section of one of the flat bones of the skull, such as the parietal, should also be prepared.

The first thing to be done is to get as thin a piece as possible cut with a fine saw from any desired part. The piece so obtained is ground down on a hone wetted with water. The hone must have been previously freed from all traces of greasy matter by washing with soap and water with a little soda added. The piece of bone is pressed down and rubbed to and fro on the hone simply by the finger, being ground first on one side and then on the other. The feel will be almost sufficient to tell when it is thin enough, and this may be confirmed by placing it on a slide without covering it and examining with a low power. Numerous scratches will doubtless be visible on its surface, produced by the grain of the stone, however fine this may be ; but unless very obvious, they may be disregarded, for they become almost invisible in the subsequent preparation. They may, if desired, be removed by polishing the section upon a piece of plate glass with the aid of putty powder. The section is well rinsed in clean water by aid of a hair pencil, placed upon a slide to dry, and, when dry, mounted in Canada balsam. The balsam to be used for this purpose must not be, as is usually the case with that sold in the shops, semi-fluid, but quite hard in the cold—a condition which results from long keeping, or may be produced by heating a little of the more recent resin in a capsule over a sand- or water-bath, until all the volatile matters are driven off. A drop of the melted balsam is then placed upon a slide, which is warmed over a flame until the resin has diffused itself pretty evenly over the central part of the slide. This is then placed on the table, and whilst cooling, but before the balsam has time to become quite hard again, the thin piece of bone is placed upon it. A clean cover-glass is then taken up by the forceps, a drop of the balsam placed upon it also, warmed in like manner, and quickly inverted over the preparation. By this mode of proceeding the balsam fills up and renders invisible the scratches on the surfaces of the section of bone, and some of the Haversian canals may also be filled by it; but it becomes solidified before it has time to penetrate into the lacunæ and canaliculi, which remain, therefore, filled with air, and present the black appearance which is characteristic of any small cavities containing air when they are viewed under the microscope by transmitted light.

Most of the structural points with regard to bone can be seen much better in these preparations of the hard tissue than in sections of a decalcified bone. But it is important in mounting the pieces after they have been ground down to be careful that the balsam does not remain fluid long enough to have time to penetrate into the thickness of the section; if this should happen, the whole is rendered too transparent for many of the details of structure to be made out. On the whole it is perhaps advisable to purchase one or two good specimens rather than devote a large amount of time to the manufacture of what may after all turn out to be but an indifferent preparation. Sections of hard bone and of teeth are amongst the very few histological preparations which are usually better made by those who make a business of preparing microscopic specimens for sale than by the student himself.

Softened bone.—In addition to sections of hard bone which are made by grinding in the manner above described, a portion of a well-macerated bone may be placed in v. Ebner's solution (hydrochloric acid 5 grms.; sodium chloride 5 grms.; water 200 cc.; alcohol 1000 cc.) until all the earthy matter is dissolved; the piece is then steeped in 70 p.c. alcohol, containing bicarbonate of soda, to free it from the remains of the acid, and finally in stronger alcohol. Sections may be made with a microtome and mounted in glycerine or, after staining, in balsam.

It is to be remembered that in considering the structure of a bone we have to deal not merely with the bony matter pure and simple and its included cavities, but that there are in addition the soft contents of those cavities—the corpuscles in the lacunæ, the blood-vessels in the Haversian canals, and the marrow in the medullary cavity and in the interspaces of the spongy tissue. In order properly to view these structures (which are destroyed in a macerated bone) we must employ a reagent which will at the same time decalcify and soften the hard matter and preserve and harden the included soft tissues.[1] Such a reagent is to be found in a solution of chromic acid, but as the decalcifying powers of this are but feeble, the portions of bone placed in it must be as small as possible, and a large proportionate amount of the fluid must be used. For example, if the bone is a long bone about the size of the human metacarpal, a piece not longer than a quarter of an inch should be sawn from it and placed (suspended by a thread) in a beaker capable of containing some 200 c.c. of the chromic acid solution. If the bone is larger, it is well to

[1] v. Koch has devised a method of obtaining by grinding thin sections of bones, teeth and the like with the soft parts preserved *in situ*, and without the necessity for decalcifying. The method is largely used for studying tooth structure, and is described later on (p. 213).

split the disc thus sawn off into three or four smaller pieces, since otherwise the decalcification will occupy too long a time. The acid used should be at first very weak (1 in 500) ; in two days' time this may be changed for a solution of 1 in 200 ; and this again in another two days for 1 in 100. Beyond this the strength of the acid should not be increased, but the fluid should be renewed every three days at least. In addition to this frequent changing of the liquid, it should, throughout the whole time of softening, be stirred as often as possible : this is of importance, for every agitation brings fresh portions of acid to attack the earthy matter of the bone. By attention to this particular the time which the pieces take to become thoroughly decalcified may be materially shortened. The completion of the process is ascertained by passing a needle through the middle of the piece employed ; if it meets with no gritty obstruction, all the earthy matter is got rid of. Should the pieces of bone be thick the decalcification process may be hastened by adding to the chromic acid solution a little nitric acid (2 c.c. to each 100 c.c. of chromic solution). It is better not to employ this reagent at the commencement of the operation, but it may be employed during the later stages, especially in conjunction with the chromic acid, since the latter reagent has then, by its coagulant action upon the tissue, so altered it that it is no longer swollen by the dilute mineral acid.

The process of decalcification being completed, the pieces are placed in running water for a day or two to get rid of the excess of acid imbibed, and are then transferred to spirit frequently changed ; or sections may be made at once without placing the piece in spirit at all. The sections should be very thin ; they are stained in hæmalum, carmalum, or magenta, washed in water, and finally mounted in diluted glycerine. The corpuscles within the lacunæ, at least their nuclei, are beautifully shown, and all the soft parts are more or less stained, but the actual substance of the bone is only slightly coloured. At the same time, any lines of demarcation which may be present indicating successive deposits of osseous matter in the formation of the bone, or absorption at any part and subsequent re-deposition, with the characteristic scolloped edge which such a junction almost always possesses, are very clearly shown by a difference in the colouration. Moreover, if in the section there happen to be any portions remaining of the (ossified) matrix of the original embryonic cartilage, this, like cartilage matrix generally, is intensely stained by logwood.

Instead of chromic, picric acid may be used for the decalcification. A saturated solution of the acid is employed, and care is to be taken constantly to supply fresh crystals of picric acid to take the place of that which is used up in dissolving the lime salts. In all other respects the proceeding is much the same as for the chromic acid method, except that the pieces will require much longer washing. This is to be done at first in 75 p.c. alcohol containing lithium carbonate, and afterwards in many changes of alcohol. The sections may be stained with carmalum or with picrocarmine and mounted in glycerine or balsam.

Two other methods for decalcifying bone, and at the same time preserving the soft tissues, may be mentioned. The first, recommended by Thoma, consists in placing the bone first in alcohol for a few days, then in alcohol containing 1 part in 6 of hydrochloric or nitric acid, and, when softened, again in alcohol containing precipitated calcium carbonate, for two weeks. The other consists in placing the tissue direct into a solution made by dissolving, with the aid of heat, 1 grm. of phloroglucin in 10 c.c. nitric acid, and filling up to 100 c.c. with water. When decalcified the tissue is transferred to alcohol. The decalcification is very rapid by the last method (a few hours), and the tissues stain well after both these methods.

Lamellæ and Sharpey's perforating fibres.—For showing these structures macerated bones that have been decalcified by hydrochloric or nitric acid serve very well. They should before use soak for some time in 10 per cent. chloride of sodium solution, which is occasionally changed to get rid of the last traces of the acid. The point of one of the blades of a sharp pair of forceps is then inserted obliquely into the bone at its outer surface, and a small piece is gripped by the other blade and dragged off in such a manner that it pulls away with it a very thin strip from the superficies of the bone. A few such strips having been obtained, either from different parts of the bone or from the same place (at different depths, therefore), they are placed in water on a slide with the inner surface uppermost and examined with a low power. In some places tapering fibres will be seen projecting from the surface of the torn-off strip like nails projecting from a board ; in other parts, round or ovalish holes, corresponding generally in size with these fibres, will be apparent : these are apertures in the lamellæ where the latter have been pierced by the fibres of Sharpey, but those fibres have been pulled out in tearing off the strips. Further, there may be made out a faint appearance of decussation in the lamellæ, like the checks on a plaid, but oblique in direction, best marked near the thin edge of the strip of bone. The strips may be mounted in dilute carbolic acid, the edges of the cover-glass being fixed by melted paraffin and then by one or two layers of gold size.

The bones in which the fibres of Sharpey may best be demonstrated in the way above described are the flat bones of the skull. They are also to be seen both in shreds and in sections taken from the long bones. If a section is made of such a bone at the place of insertion into it of a tendon or ligament, and in the direction of the fibres of the tendon or ligament, it will be seen that the bundles of fibres of the last-named structures are continued into the substance of the bone as perforating fibres or fibres of Sharpey, so that these in fact almost compose the whole of the osseous tissue at this place (Ranvier). This shows, moreover, that the fibres of Sharpey are to be regarded as bundles of fibrous tissue (connected either with the periosteum or with a tendon or ligament) which were intercalated with the osseous substance proper when this was formed and have become ossified at the same time. When tendons undergo ossification, the bony substance which is formed is wholly of the same nature as the fibres of Sharpey; this may be characteristically seen in the ossified tendons which are met with in the legs of birds.[1]

[1] For a concise account of methods which are used in the histological investigation of bone see J. Schaffer, *Zeitschr. f. wiss. Mikr.*, Bd. x. p. 167, 1893.

Development of bone in membrane.—For the study of the intra-membranous process of ossification it is best to employ the flat bones of the skull of sheep's embryos from two to three inches long. The embryos may have been preserved in bichromate of potash (2 per cent.), or spirit, or they may be employed fresh. A piece (corresponding in position with the future parietal, for example) of the still membranous skull-cap is cut out with fine scissors, placed under water if from bichromate or spirit, under salt solution if recent, and the skin and muscular layers are torn away from the outside, and the dura mater and cartilaginous layer (which in these animals rises up laterally from the cartilaginous basis cranii) from the inside. The membrane in which the bone is being formed is then left. It is held upon a glass slide with a needle or fine forceps, and brushed firmly with a camel-hair pencil the hairs of which have been cut off short so as to render the stump stiff and resisting. The piece must be kept wet, and examined from time to time with a low power to see whether the edges of the newly-formed bone are sufficiently clear of the membrane and corpuscles, so that the osseous spicules and their fibrous prolongations are readily seen. When this is the case the piece is held with the forceps and well rinsed in water or salt solution to free it from loose particles of the soft tissue ; it is then placed in carmalum solution, and when sufficiently coloured the preparation is mounted in glycerine. The osteoblasts will be found stained by the dye ; the ossified part is dark and highly refracting, the osteogenic fibres by which it is prolonged remain clear.

The process of intra-membranous development may also be very advantageously studied in sections of the lower jaw of the fœtus or young animal, as will be afterwards pointed out in describing the mode of preparing this part to show the development of other structures, such as the teeth and hair (see p. 216).

Development of bone in cartilage.—To study the later stages of ossification in cartilage it is not necessary to have recourse to the bones of a fœtal animal, since any which are still in process of growth will serve the same end, and it is more convenient for purposes of manipulation if they have attained a certain size. The long bones of a new-born kitten may advantageously be employed, and very instructive preparations may be obtained from the recently killed animal, although for permanent preparations it is necessary to put them into a fixative and hardening fluid. The mode of preparing the *fresh* specimens is as follows:—One of the long bones—the femur, for instance—is removed, and its cartilaginous head, having been cleared of the adherent soft structures, is split down the middle with a strong scalpel, the split extending a little distance into the subjacent bone. Then by a movement of the scalpel one of the halves is broken away, and the junction of the bone and cartilage is brought clearly into view. One or two slices as thin as possible are now taken from the surface thus produced, including the line of advancing ossification, and are placed on a slide in salt solution and covered. These sections are of course parallel with the long axis of the bone; but to complete the observation others should be made across that axis, at and a little below the level of the line of ossification, the cartilaginous head being first sliced gradually away until the line of ossification is reached, and then a series of sections taken of the part of the cartilage which is undergoing ossification and of the newly-formed bone.

Fresh preparations thus made are both very beautiful and instructive if the sections are obtained sufficiently thin. If it is wished permanently to preserve a section made in this way, the salt solution in which it is mounted is replaced by a solution of osmic acid (1 per cent.). The preparation may be left in this for an hour, after which time the osmic acid is rinsed away by water, and finally the sections are mounted in glycerine.

Sections, longitudinal and transverse, of ossifying bones which have been in bichromate of potash (2 per cent.) for a few days are prepared in exactly the same manner, except that the treatment with osmic acid is here unnecessary. The preparations can be mounted at once in glycerine.

To trace all the steps of the ossification, it is requisite to obtain sections of bones in various stages of formation. For this purpose the entire limb of a small fœtus is decalcified in chromic or picric or other acid, in the same way as has been recommended for the decalcification of pieces of the fully developed bone. If the fœtal bones are still small, the time necessary for such decalcification will be very much shorter than was requisite for the dense adult bone, nor need the strength of the chromic acid solution exceed 1 in 500. After completing the hardening process with spirit the limb is stained in bulk by placing it overnight in 0·5 per cent. solution of magenta in spirit. From the alcoholic stain it is transferred to clove oil and thence to xylol. It is then soaked with paraffin, and sections are cut parallel to the longitudinal axis of the limb, the sections being fixed to the slide and mounted in xylol balsam in the usual way. Other sections are made from another limb which is embedded in such a way as to be cut across the axis. Nearly all stages of ossification are found in these preparations, for the higher limb bones are usually far advanced in process of formation whilst the phalanges are only commencing ; the other bones being generally in intermediate stages. Magenta produces a double colouration in developing bone, the calcified cartilage being stained darkly, the periosteal bone of a brighter tint. The disadvantage of magenta staining is its lack of permanence. It may be replaced by a double bulk staining, first with solution of carmine or eosine, and afterwards with hæmatoxylin, any excess of the latter being removed by treatment with acid alcohol (see p. 22). The carmine-hæmatoxylin stain has the advantage of permanence.

Medullary tissue.—Ordinary marrow, which consists for the most part of adipose tissue, is obtained from the long bones of most animals. But the spongy tissue of the bones generally and the medullary canal of the long bones in some animals—the rabbit and guinea-pig, for example—are filled with *red* marrow, which contains little adipose tissue, but is mainly made up of the so-called *proper marrow-cells*, which are in many respects similar to the pale corpuscles of the blood and, like these, exhibit amœboid movements. To see the marrow-cells, therefore, in their natural condition the tissue should be taken quite fresh and examined on the warm stage. The bone—the femur of a guinea-pig, for instance, preferably a young animal—having been removed and cleared of the surrounding soft parts, is broken across, and a small piece of the marrow picked out and broken up with needles in a drop of serum or salt solution on a piece of thin glass. A small piece of hair is then added, and the preparation is covered : a brush dipped in oil is drawn round the edge of the cover-glass to prevent evaporation, and the specimen is then placed on the warm stage (p. 53) and examined. The different sizes and varying forms of the marrow-cells are to be noted, as also the large clear nucleus which many of them possess, and which at once distinguishes these cells from the ordinary white blood-corpuscles. But many are also observable which are in every respect similar to the latter, and may indeed not improbably belong to the blood, for the blood-vessels are very large and numerous in the medullary tissue, and yield the numerous red blood-corpuscles which are seen scattered over the field. Besides these colourless cells of the marrow, there are in red marrow a number of similar but rather smaller elements which resemble the marrow-cells in most respects, but differ from them in possessing a nucleus. These are the *erythroblasts*, which are characteristic of red marrow. They are in fact nucleated coloured blood-corpuscles, and are believed to give rise to the ordinary blood-discs by losing their nuclei and becoming modified in shape.

Other cells may occasionally be met with much fewer in number, flattened in form, and sometimes branched, with a large clear oval nucleus, and in some instances containing yellowish-red pigment granules. They are larger than the proper marrow-cells, and exhibit no changes of form. There is yet another element to be found in the marrow—most likely to be met with in a bit taken from near the inner surface of the bone. This is the *myeloplaxe*, and is characterised by its enormous size—whence the name *giant-cell*—by its granular appearance, and by containing a number of clear round or oval nuclei grouped together in the middle of the cell, or, in many cases, a single large irregular nucleus with numerous buds from its circumference.

To isolate these various elements of the marrow better than can be done in the fresh condition, a piece of the tissue is to be placed in one-third spirit for a day or two. After this time a small portion is thoroughly broken up with needles in a drop of water on a slide, is then covered, and stained by allowing dilute logwood solution to flow in under the edge of the cover-glass. The logwood is replaced in a minute or two by water, and this again by glycerine. By this means a permanent preparation is obtained, which can be studied at leisure, and which very well exhibits the different kinds of cells above enumerated, whilst the red blood-discs are rendered almost completely invisible. For class purposes the method recommended on p. 81 may be adopted, but with the substitution of one-third alcohol for Flemming's fluid.

Sections of marrow, both yellow and red, should also be prepared. They are made either from alcohol or from corrosive sublimate preparations (pp. 17 and 18), and the pieces, after hardening, may be stained in bulk with carmalum, followed by alcohol holding picric acid in solution, or by Heidenhain's method (p. 20). For the study of the development of red blood-corpuscles from erythroblasts of the marrow it is best to make thin sections from corrosive sublimate-hardened marrow of birds, and to stain with hæmatoxylin and eosin or carmalum and picric acid.

Cover-glass preparations of red marrow, made like those of blood (p. 80), except that the marrow is smeared over the cover-glass, also give valuable results, especially for the study of the granules of the marrow-cells.

CHAPTER VI

MUSCULAR TISSUE

Involuntary muscle.—It is easy to isolate the lanceolate cells of which this tissue consists. For this purpose all that is necessary is to place a piece of any organ containing plain muscular tissue—the intestine, for instance—in a weak solution of bichromate of potash (one part to 800 of water) or in one-third alcohol for 48 hours. At the end of this period of maceration, a small strip of either the longitudinal or the circular muscular fibres is torn off with forceps, held in a drop of water on a slide, and frayed out as finely as possible with a needle. A cover-glass is then laid on, and the preparation is carefully examined with a high power. The ends and edges of the larger pieces of the tissue have a somewhat ragged aspect, due to the projection from them of the tapering ends of the fibre-cells. In addition to these partially separated elements, others are to be met with scattered over the preparation which are wholly free and in which all the characteristic appearances of this tissue can be distinctly made out. The elongated nucleus in the middle of each riband-shaped cell can be seen in those cells which lie flat, but it is at present indistinct. It may, however, be brought into view, as can also the faint longitudinal striation which the cells exhibit, by employing as small a hole as possible in the diaphragm of the microscope to admit the light to the object. But to show clearly the nuclei of the plain muscular fibre-cells, nothing is better adapted than staining the tissue with a weak solution of logwood. It must be used quite dilute, and suffered gradually to diffuse itself under the cover-glass from a small drop placed at one edge. It is not a good plan to draw it through by means of blotting-paper, since in this way many of the isolated elements of the tissue will be drawn away at the same time. But after the drop of logwood solution has passed in great measure or entirely under the cover-glass, a small drop of glycerine may be added at the same spot as the logwood. This as it diffuses under the cover will gradually push, as it were, the logwood solution before it, so as to cause the staining fluid to traverse successively every part of the preparation, and eventually to become collected entirely at the opposite edge, the water meanwhile evaporating and leaving the glycerine in possession of the field. All that is needed to complete the preparation is, in the course of a day or two, to fix the cover-glass by painting a little gold size around the edges. Another method which yields good results is to place a piece of the tissue in picrocarmine solution and leave it in this for some days or even weeks. The fibres are then readily isolated and will be stained yellow by the picric acid, while their nuclei have assumed the red colour of the carmine.

The involuntary muscular fibres will be seen in section in preparations of the stomach and intestine and numerous other organs, so that it is not necessary to make special preparations at this stage for the purpose of thus showing them. It may, however, be instructive to demonstrate the manner in which the cells are applied edge to edge in order to make up the bundles and lamellæ of the tissue. This is often shown sufficiently well in a thin strip which has been stained with logwood and mounted in glycerine without teasing.

But the best preparations for exhibiting the arrangement of the cells are those stained with nitrate of silver. As in other tissues, this reagent stains only the intercellular substance, leaving the cells themselves uncoloured; their outlines are thus brought very distinctly into view. They may be demonstrated in the frog's bladder by the following method :—Cut out a piece of the bladder and lay it upon a slide with the inner surface uppermost. Cover it with water, and with the finger forcibly rub off the epithelial lining, washing away all débris with a stream of distilled water. Now cover with 1 per cent. silver nitrate solution and leave this on for two minutes. Wash off with distilled water, and place the piece of bladder, still attached to the slide, in a dish of spirit and expose to sunlight. When brown, which in direct sunlight will be in a few minutes, transfer the slide to clove-oil, and, when this has cleared up the preparation, dry off the excess by pressing blotting-paper upon it and mount in xylol balsam. Or the preparation after being removed from the spirit may be allowed to dry on the slide, and when completely dry may be mounted directly in xylol balsam.

Voluntary muscular tissue : cross-striped muscle.—For the examination of this tissue in mammals a small longitudinal shred is torn or snipped off from any muscle of the limbs or trunk of a recently killed animal, placed upon a dry slide, and the fibres are slowly and carefully separated from one another, one by one, for as great a length as possible, care being taken to keep them moist all the time by the breath. The cover-glass is then laid on, having first had a drop of serum or salt solution placed upon it, and the preparation examined first with a low power, to study the shape and extent of the fibres, afterwards very carefully with an ordinary high power, and eventually with as high a power as it is possible to obtain. In this preparation the following points may be made out : (1) the delicate sarcolemma ; (2) the muscle nuclei immediately beneath it, which look clear and oval when the upper surface of a fibre is exactly focussed, fusiform when seen at the edge ; (3) the dark cross-stripes of the muscular substance, seen by careful observation with a very high power to be pervaded by fine parallel longitudinal lines ending in enlargements within the clear stripes, or in a certain focus appearing to end in a dotted line in the middle of the clear stripes.

The structure of vertebrate muscle may also be studied in sections of alcohol-hardened muscles, the sections being cut both longitudinally and transversely, stained with hæmalum or picrocarmine, and mounted in xylol balsam. In these sections also the 'muscle-spindles' will occasionally be met with and may be studied.

The sartorius muscle of the new-born infant furnishes an especially favourable object for the study of those interesting structures; but they are to be found in most muscles, although in some, *e.g.*, those attached to the globe of the eye, they have not been detected. They appear to be composed of a bundle of incompletely developed muscular fibres, enclosed in a sort of capsule of connective tissue containing lymph-spaces, and having medullated nerve-fibres, afferent in nature, ramifying amongst them (Sherrington, *Journ. Physiol.* xvii. 237).

Nuclei.—For the purpose of bringing these more distinctly into view another piece of fresh muscle may be prepared as just described and mounted in dilute acetic acid (1 part in 200 of salt solution). It will be found that whereas in the former preparation the muscle-nuclei could only be made out by exercising the greatest care and attention, they are now extremely obvious, studding the fibres at intervals, but in most fibres lying at the surface of the muscle under the sarcolemma. If a frog's muscles are prepared in like manner with acetic acid the nuclei are, on the contrary, seen for the most part to be embedded in the thickness of the fibre, and this is also the case with some of the fibres from the mammal (red muscles of the rabbit).

Sarcolemma.—This is extremely delicate in mammalian muscle, and although it may with care be made out, is nevertheless much more easily demonstrated in the muscles of the lower vertebrata—the frog, for example. With this end a piece of fresh and living muscle is carefully separated as above described and covered in a drop of salt solution. A process of disintegration generally begins at places where the fibres have been touched by the needles in the process of separation, and if the muscle is living the contractile substance breaks and shrinks away at these places, leaving clear sarcolemma bridging across the interval.

Discs and muscle-columns.—In order to exhibit the manner in which muscular tissue tends to break up into either discs or columns, according to the nature of the reagent to the action of which it is submitted, two pieces of muscle are taken from an animal that has been dead some hours, and are placed for a day or two, the one in a solution of hydrochloric acid (1 in 500), the other in a solution of osmic acid (1 in 200). A small portion of each is then broken up as finely as possible with needles. The fibres from the hydrochloric acid are, many of them, found to cleave into transverse clear discs, some of which will be noticed lying flat, others seen edgeways; whereas in those from the osmic acid there is no such tendency to form discs, but, on the contrary, the muscular fibres tend to break up into longitudinal columns.

Isolation of fibres.—In order the better to compare the fibres either of the same or of different muscles as regards length and diameter, and to see their general shape, it is necessary to isolate a number of them in their whole length. For this purpose the process of separation by the aid of needles is somewhat tedious, and we must turn to reagents which will dissolve the intermediate connective tissue which binds the fibres together, whilst maintaining them intact. Such a reagent is to be found in a solution of sulphurous acid (liquor acidi sulphurosi of the Pharmacopœia). The muscle is placed for a week or more in a well-stoppered bottle containing a considerable quantity of the acid, and is kept in a warm chamber heated to about 40° C. This facilitates the process of maceration, so that after the time stated a mere gentle shaking of the bottle is sufficient to cause the muscle to break up in great measure into its constituent fibres. Some of these may then be removed, placed side by side in water or weak glycerine on a slide, and covered, with the usual precautions to obviate the pressure of the cover-glass. It will be found that the muscular substance has acquired, in consequence of the maceration, a granular aspect, and that the usual structural appearances are for the most part indistinct. But the mode of preparation may be nevertheless employed for purposes of measurement and comparison of size and form of fibres from different regions of the body.

Study of insect muscle.—Many insects possess two kinds of striped muscular tissue, the one greyish looking and composed of large fibres having the same appearance under the microscope as vertebrate muscle, except that the characters are more distinct, the other of a yellowish aspect, also composed of large fibres, which have a less marked cross striation than the others, but readily break up into longitudinal elements (sarcostyles), which are themselves, however, markedly striated. Of the latter kind are constituted the muscles which move the wings; of the former the muscles of the trunk and legs. They can be studied both in the living condition and after preservation in alcohol.

Examination of muscular tissue of insects in the living condition.—One of the most convenient insects to employ for the purpose is the common great water-beetle (*Dyticus marginalis*), but almost any beetle or other insect (especially the wasp) may be employed instead. The head is first cut off and the trunk is then split longitudinally with scissors.

When this has been done the two kinds of muscular tissue will be noticed, and a portion of each can be taken up with the point of a scalpel, rapidly teased on a slide in a drop of white of egg, and covered. The preparation is to be examined at once with the highest available power. It is difficult to make out the details of the structure with a combination magnifying less than 1,000 diameters, and the defining power must be of the best.

In the limb muscles, if the object has been quickly enough prepared, numerous fibres will be found which show the characteristic successive series of minute longitudinal lines of interstitial substance running through the darker cross-stripes of the muscle, and prolonged into the clear stripes, where they pass into enlargements, the juxtaposition of which in rows side by side gives the semblance of a dotted line in each clear stripe, single or double according as the fibre is more or less extended. Other fibres may be seen in which the dotted line is absent, the longitudinal lines not being enlarged within the clear stripes ; and these show especially well the muscle-columns which extend in the direction of the length of the fibre. In such fibres the cross-striped appearance is indistinct, due to the fact that the clear striæ are but slightly marked, their bright appearance in the other fibres being greatly enhanced by the presence of the rows of strongly refracting dotlike enlargements of interstitial substance.

In the wing muscles many of the muscle-columns will be isolated, and in them it is possible to make out the transverse membranes of Krause in the middle of each clear stripe and in the more extended sarcostyles, the line of Hensen bisecting each dark stripe.

In the perfectly fresh preparations of the insect's muscular tissue, spontaneous waves of contraction may be seen passing from end to end of many of the muscular fibres, and a general idea of the phenomena which accompany the contraction, such as thickening of the part of the fibre at the time the wave is passing, and approximation of the cross-striæ, may be obtained. But the contraction proceeds too quickly for all the details of the process to be followed.

Preparations from alcohol-hardened muscle.—The structure of muscle can for the most part be better made out in alcohol-hardened preparations than in those fixed by any other method, and the alcohol method has the advantage that the tissue can be stained by various reagents prior to being mounted. One only of the modes of staining need, however, here be given, viz., the gold method of Rollett, which is a special stain for this tissue.

It will be best, before proceeding to this, to examine

teased preparations of alcohol-hardened insect muscle without staining. To fix the muscles, the insects are thrown alive into strong spirit, in which they are left for at least a day. They are then split open longitudinally, and portions of the two kinds of muscle removed and teased (*a*) in dilute glycerine, (*b*) in vinegar. In the former the sarcostyles are prominently seen ; but in the vinegar preparation the sarcostyles become swollen, and the interstitial substance (sarcoplasm) comes distinctly into view with its appearances of lines and dots. In this preparation also there will be a tendency for the fibres to break across into discs, and these discs, when seen on the flat, exhibit very well the appearances of the fibres as seen in cross-section.

Rollett's gold chloride method of staining muscular tissue.—A small piece of muscle, from either the ordinary or the wing-muscle of an insect which has been in 90 per cent. alcohol for from twenty-four to forty-eight hours, is steeped for three or four hours or more in strong glycerine. From this it is transferred to 1 per cent. chloride of gold solution, and allowed to remain for from fifteen to thirty minutes. It is then placed in formic acid (1 part formic acid to 3 parts water), and kept in the dark for twenty-four hours or more, after which it may be teased in glycerine. The fibres which have been prepared by this method show many varieties of staining, probably dependent upon the variable time they have been exposed to the chloride of gold (for of course the fibres near the surface are longer exposed to the reagent than those nearer the middle of the piece of muscle), and some will be found the interstitial substance of which is stained, and has the appearance of a fine network throughout the muscle, others in which this remains colourless while the sarcostyles are stained. In the latter, it will be found that it is the substance of the dark disc (sarcous substance) in which the metal is deposited, so that the sarcous elements have a red or purple colour, while the substance of the clear disc is left completely colourless. In these preparations of alcohol-hardened muscle, fibres and sarcostyles will be found in varying conditions of extension and contraction, so that the phases which mark the passage of a fibre from the state of full extension to that of full contraction can be observed.

Examination of muscular tissue by polarised light.—The *polarising microscope* is nothing else than the ordinary microscope with the addition of two Nichol's prisms, one placed below the object and another above the ocular ; the upper one is generally mounted in combination with a low ocular, so that it is not necessary to use the ordinary eye-piece. The light, coming from the mirror, becomes polarised as it passes through the lower Nichol (the polariser). If now the upper Nichol (the analyser) be slowly turned round as it is being looked through, it will be found that there are two positions in which the field is quite dark ; that is to say, the polarised rays are entirely cut off. By observing now the relations of the prisms at these positions of total darkness, it will be found that their planes of polarisation—as shown by the way in which the prisms are cut—are at right angles to one another. In all intermediate positions a greater or less amount of light is enabled to traverse the analyser. But if any object which possesses the property of refracting light doubly is placed upon the stage of the microscope, and examined. and if then the field is made dark by turning the analyser, it will be found that the doubly refracting substance remains bright, unless it happen so to lie that its optic axis is parallel with the plane of polarisation of either Nichol.

The observations may be made upon the living or upon alcohol-hardened muscle. The portion of fibre under observation should be quite free and not overlaid by other fibres. The change in the optical condition of the fibres which ensues on contraction may, if due care and patience be exercised, be made out. The results arrived at by the examination of portions of the tissue which have been hardened in alcohol, and mounted in glycerine or Canada balsam, are more easily seen than if living muscle be examined.

A modification may be made by substituting a thin piece of mica for the covering glass. This causes the field of view to become tinted, the particular colour varying with the thickness of the mica and the relative position of its optic axis to those of the Nichols, and any doubly refracting substance which is now examined assumes the colour which is complementary to that of the field. The object of the revolving stages with which the larger microscopes are generally fitted is to enable the observer to modify the position of the optic axis of the tissue which is being examined with relation to those of the Nichols; and it also serves when the mica is used to change in like manner the relative position of the optic axis of this also, and thus to modify the colour of the field of view.

Transversely striated muscle is not the only tissue which is doubly refracting, for the property is possessed by the white fibrils of connective tissue, and by bone, as well as by plain muscular fibre-cells, and to a less degree by ciliated epithelium. But it is the only one which exhibits alternate bands of singly and of doubly refracting substance. It has, however, been pointed out by Ranvier that it is rather the conditions of tension of a tissue than differences of structure which tend to determine differences in the optical properties of the substance of which it may be composed. Thus he instances the case of cartilage, the matrix of which, although undoubtedly composed of the same substance throughout, becomes doubly refracting in those parts where the cells, either from pressure or in progress of growth, assume a flattened or elongated shape, singly refracting where they remain rounded.

Transverse sections of insect muscle.—Fairly good views of the sectional appearances of the substance of a muscle-fibre may be accidentally got in the teased preparations which have already been described. But to obtain complete sectional views of the muscles pieces of hardened muscle are embedded in paraffin, cut very thinly with a microtome, and mounted by the adhesive method. Both the ordinary and the wing muscles of insects are to be prepared in this way : the pieces of muscle may be stained in bulk previous to embedding, or on the slide after section.

Termination of muscle in tendon.—If the tendons of the tail of a mouse or rat are forcibly drawn out, after nipping off the end of the tail in the manner described in the account of the preparation of tendon (p. 108), it will generally be found that portions of a number of small muscular fibres are adherent to each tendon, for the fibres have their insertions into the tendon, and are ruptured by the force employed. These ends, mounted in serum, serve conveniently for the study of the mode in which the fibres of a muscle terminate in a tendon when the fibres of the latter run in the same direction as those of the muscle. But, easy as the tissue is to prepare, the observation is complicated by the fact that the muscular fibres form generally somewhat of a clump as they pass to end in the tendon. The preparation may be improved by being stained with picrocarmine solution. This colours muscular tissue yellow, tendinous tissue red, so that the distinction between the two is made more obvious. It is best to take a few freshly drawn-out tendons, and to mount their ends in a drop of the picrocarmine solution, surrounding the edges of the cover-glass with melted paraffin to prevent evaporation of the liquid.

Ranvier's method.—A frog is pithed and is placed in a litre of water at 55° C., which is then allowed to cool. After a quarter of an hour the animal is removed from the water, the skin detached from one of the limbs and a small piece of muscle taken, including a piece of its tendon, and carefully dissociated with needles in a drop of picrocarmine solution or in serum to which iodine has been added. It will be found that the muscular substance has for the most part retracted from the ends of the sarcolemmal sheaths, and that in favourable cases the attachment of the small tendon bundles to the ends of these sheaths can be made out without much difficulty.

MUSCLE 145

The attachment of muscle to tendon may also be studied in sections of a muscle passing through its tendinous attachment.

The blood-vessels of muscle will be studied later, in sections of the injected tissue, after the methods of injecting have been described. The mode of termination of nerves in voluntary muscle will also be deferred until the preparation of the nerves themselves has been treated of.

CHAPTER VII

NERVOUS TISSUE

Medullated nerve-fibres.—For the study of medullated nerve-fibres a piece of one of the ordinary nerves—those of the limbs, for example—may be cut out from any recently killed animal. If the nerve is a large one, a thin strip only should be used, preferably taken from the interior of a funiculus torn longitudinally into two pieces by fine forceps. The strip is to be placed on a slide in a little serum or salt solution, and carefully separated as finely as possible. This separation must be effected, not by seizing the piece anywhere and tearing it up at random, but by inserting fine needles into it near one end and gently drawing them asunder, so that the piece is split into two. Repeating this process a number of times on the resulting pieces, the nerve will be eventually separated into very fine bundles of fibres, together with a number of more or less isolated fibres, which are still nearly straight and uninjured, except near one end. The preparation may then be covered, and the general character and appearance of the fibres investigated. To see the nodes of Ranvier well, a tolerably large fibre, free for a considerable part of its length, should be chosen, and by moving the slide it should be carefully followed with an ordinary high power. It will be found that at definite and not very close intervals along the fibre the double-contoured medullary sheath fails altogether, and the axis-cylinder alone appears to continue the nerve at these points. It is not easy to see the oval nucleus in the middle of each segment in the fresh, unstained preparation, but the obliquely truncated segments into which the medullary sheath tends to split are very obvious.

Preparation with osmic acid.—Osmic acid possesses the property of staining the medullary sheath of the nerves, which is mainly composed of lecithin, of an inky-black colour, whilst the other parts are left of a greyish tinge. Moreover, the breaks in the fatty sheath—both those at the nodes of Ranvier and the irregular breaks between the medullary segments—are by this means brought into prominence, owing to the intermission of the darkly stained medullary substance at those places. At the same time the axis-cylinder can be distinctly made out crossing the intervals, and the primitive sheath, or sheath of Schwann, can be seen. The method of preparing these osmic preparations is as follows :—From an animal which has been quite recently killed, a small nerve, or piece of nerve-root, not larger in diameter than an ordinary thread, is chosen, and a piece about half an inch long is cut out, but in doing so care must be taken not to drag upon or injure the nerve more than is absolutely necessary. The piece is placed for four hours in a small covered glass pot containing a few drops of 1 per cent. solution of osmic acid; it is then transferred to water for an hour, and finally placed in a mixture of glycerine and water (equal parts). It can either be teased in this at once or left for a day or two or even longer; in preparing it, the same precautions must be used as were recommended for the preparation of the fresh nerve (see preceding paragraph).

Fibres of Remak.—To see the grey or non-medullated fibres, pieces of the sympathetic nerve are taken from the neck of an animal and prepared and examined both fresh and after treatment with osmic acid, in the same manner as the cerebro-spinal nerves. Many small medullated fibres are here, however, found intermingled with the non-medullated. To obtain non-medullated nerves almost entirely free from medullated it is best to make preparations from the splenic nerves, which can be readily got at, and in some animals are comparatively large. Non-medullated fibres should be fixed as rapidly as possible after death, as they quickly undergo structural changes.

Mode of union of the nerve-fibres to form the nervous cords.—In the several teased preparations both of the cerebro-spinal and of the sympathetic nerves there will be seen, especially in those prepared with osmic acid, besides the actual nerve-fibres, in the first place, a quantity of connective tissue, for the most part of the nature of areolar tissue, which formed a general ensheathment for the nerve, and sends partitions in between its several bundles or funiculi; secondly, the special sheaths of the funiculi, which become torn and stripped away in the process of teasing, and which look like flat bands composed of an almost homogeneous substance, but are pervaded by a network of fine fibres and with round or oval nuclei scattered upon them here and there. Thirdly, there will be seen running along close to and surrounding the nerve-fibres themselves very delicate, nearly straight fibrils of connective tissue, with here and there the nucleus of a connective tissue corpuscle. These three forms of connective tissue represent respectively the epineurium, or outer sheath; the perineurium, or funicular sheath; and the endoneurium, or tissue within the funiculus. Their relative situation and arrangement, as well as the lamellated structure of the perineurium, can only be properly displayed by transverse sections of a nerve-trunk.

But the cell-outlines on the lamellæ of the perineurium may be shown by the silver method. For this purpose either a very small nerve is chosen, and a piece of it is removed and immersed for five minutes in half per cent. silver nitrate solution, after which it is washed in distilled water and exposed in glycerine to sunlight, or, should it be wished to prepare a larger nerve consisting of more than one funiculus, this is partially dissociated in a few drops of the silver solution, and after a like treatment is also mounted in glycerine and exposed to the light. After a few minutes' exposure to sunlight the preparations may be examined. It will be found that the sheath of each funiculus or nervous bundle is covered by large epithelioid markings, and if the preparation is successful, two, three, or even more layers deep of such markings may be counted by examining with a high power and carefully adjusting the microscope.

In addition to this it will generally be found that any medullated nerve-fibres to which the silver solution has penetrated are marked transversely at each of the nodes of Ranvier by a dark line, or rather a perforated disc, surrounding the axis-cylinder. These markings appear to be owing to the existence of a substance here, between the segments of the nerves, which, like the intercellular substance elsewhere, has an affinity for the metal. Finally, at many of the nodes of Ranvier, particularly if the immersion in the silver solution had been rather prolonged, the latter will have penetrated as far as, and will have become reduced in the substance of the axis-cylinder, which therefore will here have the appearance of a dark cylindrical rod piercing the ring of intersegmental substance, the two together having the semblance, under a moderate power, of a little black cross upon the nerve at those points. In such preparations the axis-cylinder, where stained at the nodes by the reduced silver, has a cross-striated appearance.

Sections of a nerve-trunk.—To prepare a nerve-trunk for microscopic section, it should be hardened in picric acid, osmic acid, Flemming's fluid, or bichromate of potash. It should be gently extended upon a piece of cork, its ends being fixed with glass pins, and it is best, if the nerve is large enough, to ensure rapidity of hardening by introducing some of the hardening fluid into the interior of the nerve-trunk by a glass pipette with a capillary point. The piece is left in picric acid (saturated solution) for two days, and then transferred to spirit ; in 1 per cent. osmic acid for 24 hours, then washed with water for two hours and transferred to spirit ; in Flemming's solution for two days, and then transferred to spirit ; or in bichromate of potash for seven days, and then

transferred for three days to a mixture of 2 per cent.
bichromate potash (2 parts) and 1 per cent. osmic acid
(1 part). This piece also is placed in spirit prior to embedding.
Sections may then be cut either by the freezing or paraffin
method, and mounted either in glycerine or in xylol balsam,
but the sections, except those from the nerve hardened in
osmic acid, must first be stained with aniline blue-black,
with hæmateïn, or with carmalum.

Study of degenerating nerve-fibres.—The changes which
medullated nerve-fibres undergo in consequence of severance
from their cells of origin (which in the case of the sensory
fibres of the spinal nerves are situated in the ganglia of the
posterior roots, and in the case of the motor fibres of the
spinal nerves are placed in the grey matter of the spinal cord)
may be studied in the peripheral part of nerves which have
been cut from two days to three weeks previously, the nerves
being treated in precisely the same way, both for teased preparations and for sections, as normal nerves.

Nerve- and ganglion-cells ; neuroglia-cells.—To study
the exact form and appearance of the tissue elements of the
nervous centres they may be isolated and examined in teased
preparations, although their position and local relations are
best made out in sections of the several organs in which they
occur. For the present we will confine ourselves to a
description of the best methods for isolating the cells. For
this purpose the tissue is treated similarly, from whatever part
of the brain or spinal cord the piece to be examined is removed ;
but, as a typical example, a piece of the spinal cord from the
lumbar region may be taken, for the nerve-cells are here very
numerous, and consequently more readily found than in the
other regions.

Cells of nerve-centres.—The human spinal cord is, if
obtainable, best adapted for the study of the cells, for they
are readily picked out under the dissecting microscope, in consequence of the little mass of dark pigment which each contains. If a piece of the human spinal cord cannot be got, the

spinal marrow of the ox or calf, or some other large animal, may be employed. Small pieces of tissue are dug out from the grey matter, preferably from that of the anterior cornu, and placed in one-third alcohol. After two or three days' maceration in this, the pieces are shaken up in a test-tube with some of the fluid and allowed to subside. The supernatant fluid is then decanted off, fresh one-third alcohol added, and the shaking and decantation repeated. Some of the débris may now be transferred to a slide, a piece of hair placed in the fluid, and the cover-glass gently superposed. The cells may then be sought with a low power and carefully examined with a higher one.

If the preparation is successful, and it be desired to preserve it, keeping the cells at the same time as much as possible of their natural aspect, the best plan to adopt is to allow a drop or two of a 1 per cent. solution of osmic acid to flow in at the edge of the cover-glass, and after it has acted upon the tissue for an hour, by which time the cells will have become stained of a dark greyish tint, to carefully run through from the same side first a little distilled water to wash away what remains of the acid, and then glycerine to preserve the preparation, after which the edges of the cover-glass may be cemented. Or the cells may be stained with picrocarmine in the test-tube before being mounted. For this purpose the supernatant one-third alcohol is poured off, and picrocarmine solution added and kept in contact with the débris for several hours or days. Besides neuroglia-cells and nerve-cells, which in a carefully prepared specimen may be very well seen, with their long branching processes extending in some cases far beyond the field of the high power objective, some points in the structure of the nerve-fibres of the spinal cord can be well made out in these preparations. Thus fibres may be readily found in which the medullary sheath has broken away in many parts from the axis-cylinders, and where not actually removed has become swollen and coagulated in irregular masses around the axis-cylinder ; changes which could hardly have taken place were there any such structure surrounding the nerve-fibres as the primitive sheath of the peripheral nerves. Further, where the axis-cylinders are in this manner laid bare, as they often are for a considerable part of their length, their fibrillar structure can, with a high power, be made out without difficulty. A similar structure can also be seen in the processes of the nerve-cells, and extending from them through the body of the cell itself.

Cells of ganglia.—Cells from ganglia, whether spinal or sympathetic, are isolated in a manner similar to that employed in the case of the spinal cord, except that the period of maceration in the one-third alcohol may be longer, owing to the much larger amount of connective tissue by which the nervous elements are invested. Prolonged maceration in osmic acid (1 to 500 of water) is also a method of much value for ganglion-cells. It is possible to get a certain number of the cells sufficiently isolated, even from a fresh ganglion, without any maceration; for each cell being loosely contained in a special capsule of flattened cells, it readily falls out when the nerve fibres with which it is connected are ruptured. But although the cells themselves of the ganglia are readily enough separated, it is more difficult both in the fresh and in the macerated preparations, to show their continuation to the nerve fibres, and the T-shaped branching of these. The mode of permanently preserving the specimen is like that employed in the preceding preparation.

Study of nerve-cells by the silver-chromate method.—Nerve-cells, whether in nerve-centres or in the ganglia, may be displayed with the cell-body and all its processes of an intense black colour by the method of Golgi, modified by Cajal :—A small piece of the organ to be investigated, not larger than half a pea, and preferably taken from a developing animal,[1] is placed fresh in a considerable quantity (not less than 20 c.c.) of

---

[1] The spinal cords of older chick embryos and of new-born mice or rats are very well adapted for the purpose, and can be prepared within a piece of the vertebral column.

a bichromate osmic solution freshly made by mixing 3 volumes of 3 per cent. bichromate of potash solution with 1 volume of 1 per cent. osmic acid. The piece is left in this mixture from 1 to 5 days,[1] preferably at a temperature of 25° C. and in the dark, and is then dried with blotting-paper and transferred to a little 0·75 per cent. solution of silver nitrate. After being in this for a few minutes, the piece is placed in a larger quantity of the silver solution, to which 1 drop of formic acid is added to every 100 c.c. or 200 c.c. In this the piece is left for from 24 hours to 6 days. It is then placed in 96 per cent. alcohol for half an hour, after which it may be either placed in collodion for a few minutes, and by this fixed upon a brass holder and cut into sections with a microtome, or it may, if fairly hard and if large enough, be simply held in the fingers and sections made with a good razor wetted with 96 per cent. alcohol. The sections are passed quickly through clove-oil and xylol, and mounted without a cover-glass in thick xylol balsam or xyloldammar. As a general rule they need not be cut thin, for the greater part of the tissue remains unstained and only a few of the nerve and neuroglia cells are coloured. It is on this account that it is easy to follow the processes. It will be found that most nerve-cells have a number of processes which ramify in the neighbourhood of the cell-body (dendrons) and one process which does not ramify until it has passed a greater or less distance from the cell-body (neuron). The axis-cylinder of the medullated nerve fibres is always the neuron of a cell the nucleated body of which lies in a ganglion or other nerve centre.

Since the chromate of silver serves to stain nerve-cells and their processes and since nerve-endings are always the ramifying terminations of nerve-cell processes (neurons), it is natural to suppose that nerve-endings will also be stained by this method. It is, in fact, one of the most valuable methods

[1] According to the nature of the tissue and the structures it is wished to exhibit. For peripheral endings of nerves 6 to 8 days in the osmium-bichromate mixture is best, and it is often necessary, to get a successful result, to put the piece from silver nitrate back again into osmium-bichromate for some days, and then again in silver nitrate (duplicate process of Cajal).

for displaying nerve-endings in various parts (muscles, glands, &c.), and is employed exactly as above described. Only it must be understood that different organs require different times of immersion in the osmium-bichromate solution, and the best time for each tissue or organ can only be learned by experience. Previous hardening in formol does not interfere with Golgi's method.

Terminations of nerves.—The description of the mode of preparing and demonstrating the terminations of nerve-fibres in various special parts of the body will be deferred until those parts and organs are severally treated of, but the Pacinian bodies, in which many of the sensory fibres end, and the end-plates in which the nerves supplying the voluntary muscles terminate, may be now prepared.

The **Pacinian bodies** are very readily found in the cat's mesentery. Here they are at once seen when the abdomen is opened and the membrane is held up against the light, as clear, oval specks, either dotted singly here and there or forming groups of two, three, or more. There is generally a considerable group in the meso-rectum, and moreover they are here usually not so much obscured by the adipose tissue as in the mesentery proper. By far the best general idea of their structure and the relation they bear to the nerve-fibre entering them is obtained from their study in the fresh condition, without the addition of reagents. But it is well to separate them from the surrounding tissue of the mesentery, for this is often loaded with fat, and, when not, the fibrous tissue of the membrane tends to obscure the structure of the little bodies.

In order to isolate one of them cut out the piece of the mesentery containing it, carrying one of the cuts close along the edge of the corpuscle. Then place the excised piece upon a slide in a drop of serum, and without actually transfixing the Pacinian itself, tear away the investing mesenteric tissue bit by bit under the dissecting microscope. It will be found that with a little manipulation the corpuscle shells out as a lemon-shaped body, with a twisted stalk at one end. It is to be examined with a low power to make sure that the fat is entirely removed; the débris may then be wiped away, a little fresh serum added, a narrow slip of thick paper placed close to the corpuscle to prevent too great pressure of the cover-glass, which is then laid on and the specimen examined.

In these fresh preparations, owing to the extreme transparency of the layers of the capsule, the core of the corpuscle with the central fibre, together with the mode of passage of the nerve-fibre into this at the stalk, can be made out better than after the action of reagents. And still more so if the outer layers of the capsule are removed altogether, as may readily be done with fine needles under the dissecting microscope, leaving only the core and the closely-set layers of the capsule which immediately surround it. In tearing away the outer layers it will often happen that the perineural (Henle's) sheath which surrounds the nerve-fibre as this passes into the corpuscle is torn off along with them, since they are directly continuous with it.

These fresh preparations of the Pacinians are very beautiful, but unfortunately they cannot be preserved in that state. Treatment with glycerine causes the corpuscles to shrink and become too transparent, and most of the ordinary staining fluids colour the core too deeply and obscure the termination of the nerve-fibre. If it be desired to preserve any such preparation as showing some one or more points particularly well, the cautious employment of osmic acid prior to mounting in glycerine is most to be recommended. The serum in which the corpuscle is mounted must first be replaced by salt solution; this again by 1 per cent. solution of osmic acid, and this, after being left for an hour or more in contact with the preparation, by distilled water, whilst finally a drop of glycerine is allowed gradually to diffuse in from the edge of the cover-glass.

The structure of the tunics which form the lamellated capsule of the Pacinian body is not well shown in the fresh preparation owing to the transparency of the object; indeed, in this it is the lines of contact between successive coats which look like layers of the capsule; the substance of the coat, being clear and pellucid, gives the notion of an intermediate fluid. In order to show their fibrous connective tissue structure (the white fibrils running transversely, and collected for the most part near the surface of each tunic, and the elastic fibrils forming a network through the thickness) some of the little bodies should be dissected out in the way above described, and placed for ten days or a fortnight in a 0·2 per cent. solution of chromic acid. They are then put up on a slide in a drop of water, and with fine and perfectly clean needles are broken up under the dissecting microscope bit by bit, commencing at one end and breaking off transverse pieces. It will be found that, owing to the direction of the fibrils, the corpuscles tend to break across into disc-like portions. The core does not share this tendency, but this is of little consequence for it is not well displayed in these preparations. A piece of hair having been added, the preparation is covered and examined. Small fragments will probably be found which give a sectional view of the tunics, and others in which they are seen flat. If a little logwood solution is allowed to run under the cover-glass, nuclei on the surface of the tunics may here and there be stained.

These nuclei belong to flattened epithelioid cells, which cover both the outer and inner surface of each tunic, and which, seen in profile, are in reality the well-defined lines seen in the fresh Pacinian, and long described as the coats themselves. The outlines of these cells may be brought into view by staining with nitrate of silver. For this purpose, as is always the case with silver-preparations, the tissue must be fresh and unacted on previously by any other reagent. One or two corpuscles are to be thoroughly isolated and freed from surrounding mesenteric tissue and fat, placed for five minutes in silver solution (1 in 100), washed in distilled water, and exposed in glycerine to the sunlight until of a greyish colour, when they may be covered, and examined.

Sections of Pacinian corpuscles.—To complete the study of the Pacinians, sections should be made of them. A convenient way to prepare them, in order to show the various parts to advantage, is as follows :—A very small piece of mesentery or meso-rectum of the cat, containing several corpuscles close together, is cut out (if such can be found ; if not, one or more may be isolated as before), and placed in a small beaker containing 100 c.c. of a weak solution of acetic acid (1 in 200), to which about 5 c.c. of one-half per cent. chloride of gold solution has been added. The tissue is kept in this, in the light, for three or four days—until it has become of a dark violet colour ; it is then placed for a day in weak spirit, and then in strong, and two or three days later is ready for embedding and cutting. In embedding the piece of tissue, it should be so placed that the corpuscles, at least most of them, are cut as nearly as possible transversely.

Motor nerve-endings.—The **end-organs** (end-plates), or terminal expansions of the motor nerves, are difficult to find in fresh muscle, and so soon undergo alteration as speedily to become unrecognisable. The best muscles to choose for the search are those of a lamellar shape and with short fibres, such as the intercostals of small animals. The muscular fibres are severed close to their attachments, so as to get them in their whole length, and the small, thin piece of muscular tissue obtained is quickly transferred to a slide, and mounted, either without addition of fluid or in a small drop of serum, which is put on the cover-glass before this is inverted over the preparation.

This should now be thoroughly searched with a good high power objective for the nerve-endings. Branches of the intercostal nerve will be found running across the direction of the muscular fibres. Starting from one of these, trace carefully one by one the single nerves which pass off from it. It will be found generally that they branch one or more times, and eventually the resulting twigs pass off to the muscular fibres, each fibre receiving one of the nerve-twigs. They retain their medullary sheath until the muscular fibre to which they are attached is reached, when the sheath suddenly ceases to be visible, and it is by following the single fibres until they come in this way to an abrupt termination, that an end-organ may be met with. But even if the place where the nerve-fibre joins the muscular fibre is arrived at it is still in most cases difficult to make out the exact mode of termination, in other words, the structure of the organ. The utmost that can generally be seen is a clump of clear, round nuclei embedded in a granular material. The difficulty arises partly from the readiness with which these structures undergo alteration after removal in warm-blooded animals, and partly from the fact that they are often obscured by super- and subjacent muscular fibres or blood-vessels.

In the common lizard (*Lacerta agilis*) the motor end-organs may be much more easily found and satisfactorily seen in the fresh condition, but still the utmost care must be taken in the preparation. The animal having been decapitated and the trunk pinned out upon a cork, a piece of one of the limb-muscles—including the whole length of the fibres—is removed and placed on a slide in a drop of serum. The fibres are then separated under the dissecting microscope, carefully but not too completely, and a piece of paper or a hair having been added to avert pressure on the tissue, it is covered and the fibres are examined with a high power along their whole length. If no end-organs can be found in the first specimen another must be taken, but it will generally not be very long before one is found, either at the edge of a fibre, and therefore seen in profile, or on the surface, and seen flat.

Permanent preparations of motor nerve-endings.—Besides the method of Golgi which has been already given (p. 152), two other methods of displaying nerve-endings may be here described as being of great practical value. They are the methylene blue method of Ehrlich and the gold chloride method of Cohnheim as modified by Löwit. Of the two, the former is the more general method, since methylene blue appears to have a specific affinity for living nerve structures, and will generally, when in sufficiently dilute solution, stain these in preference to any other tissue elements. Gold chloride, on the other hand, is apt in many tissues to stain the protoplasm of cells generally, so that the nervous fibrils which may be also stained are often obscured. Nevertheless for displaying certain nerve-endings, such as those in muscle and in the cornea, there is nothing better than a gold chloride preparation. The manner in which these methods are applied to show the nerve-endings of muscle may serve as a type of their application to other tissues, although it may be advisable in some cases to adopt slight modifications in the method.

Application of the gold chloride method to motor nerve-endings.—A small piece of muscle is taken from a freshly killed animal (preferably a lizard, from the relative ease with which the fibres may be separated and the nerves followed to their endings), and placed for half to one minute in a solution of formic acid (1 part to 4 of water). It is then transferred to 1 per cent. solution of gold chloride, and after 15 minutes immersion in this it is replaced in the formic solution and left in this, in the dark, for 24 hours, after which it is put into pure formic acid for another 24 hours. The tissue should then be of a dark violet colour on the surface but redder near the centre. Small shreds are torn off with forceps, placed on a slide in glycerine and water, and slightly teased with needles, but not so as to separate the individual fibres. A cover-glass is now placed on the preparation and pressure is made upon it so that the tissue is flattened out, but without actually crushing the muscle-fibres. The larger nerve branches are now looked for with a low power and followed in their ramifications; a task easily performed if the nerve-fibres are stained—as they commonly are—of a dark violet colour, and the muscle-fibres only of a faint greyish violet.

Pieces of the stained muscle may also be cut by the freezing method. With this object they are placed first in water to remove the formic acid, then in dilute gum, frozen, cut into sections longitudinal or transverse, and the sections are mounted in glycerine or glycerine-jelly.

Application of the methylene blue method to motor nerve-endings.—Half a cubic centimetre of solution of methylene blue (1 to 100 of salt solution) is injected into the anterior abdominal vein of a frog, and the animal is killed after about an hour. Small pieces of muscle (or of any other tissue in which it is wished to show the nerve-endings) are now removed, slightly teased in salt solution and exposed to the air on a slide or in a watch-glass. After a short time they can be covered and examined, when it will be found that the nerves, even in their finest ramifications, are of an intense blue colour, the other tissues being stained little or not at all. In the case of mammals a few cubic centimetres of the methylene blue solution is injected into a vein. A modification of the method is to place the tissue to be stained, after slightly teasing it, in a dilute solution (0·2 per cent.) of methylene blue in salt solution in a watch-glass for half an hour or more, and then proceed to examine it.

It has been shown that the stain thus obtained by methylene blue can be fixed by a proper application of molybdate of ammonia (Bethe). To effect this the tissue, after removal from the methylene-blue injected animal, or from the methylene-blue solution in which it was placed, is rinsed with normal saline and then transferred to the following solution, which should be quite cold and freshly prepared:

Molybdate of ammonia	10 g.
Distilled water	100 c.c.
Hydrochloric acid	10 drops

In this it is left half an hour to an hour or more, according to size, thoroughly washed with water, and transferred to alcohol containing 0·3 per cent. platinic chloride (Cajal).

After this it can be embedded and sections cut by the usual methods.

If too strong a methylene-blue solution be used, or it be allowed to act too long on a tissue, other structures become stained besides nerve fibrils, and especially intercellular substance of connective tissue and epithelia, so that appearances like those produced by silver nitrate may be obtained.

CHAPTER VIII

THE BLOOD-VESSELS

The larger blood-vessels.—The epithelial lining (endothelium) can be best demonstrated in fresh blood-vessels by staining with nitrate of silver. For this purpose a piece of a large vessel—artery or vein—is obtained, either from a recently killed animal or from an amputated limb, and having been slit open with scissors is pinned on a cork with the inner surface uppermost. Care must be taken in doing so not to rub this surface in any way. The preparation is then rinsed with a little distilled water from a wash-bottle, with the object of removing any blood that may remain on the wall of the vessel, and a few drops of half per cent. nitrate of silver solution are allowed to flow over the surface. After a minute it is again washed with distilled water, and is then put at once into a beaker of spirit and placed in the sunlight. After a time the surface will have acquired a greyish tinge, with a browner patch here and there ; it may now be removed from the light, but should be left in spirit until the next day, when it will be hard enough to enable thin sections to be made with a razor from the inner surface. In order to cut them it will be found convenient to remove the piece of blood-vessel from the cork, and to hold it by one end by the thumb and fingers of the left hand, so that the piece rests by its outer surface on the ball of the finger ; the razor is then dipped into spirit, and as thin a slice as possible (it need not be very large) is removed from the inner surface of the blood-vessel and placed in spirit, after which one or two more may be taken from other parts. In making sections when the piece is held in this way, it will be found convenient to cut *from* the operator.

The slices are taken up on a needle or section-lifter, passed through clove-oil, and then placed on a slide with the stained surface uppermost. A drop of xylol balsam is added, and then the cover-glass. On examining with the microscope, the outlines of the epithelioid cells will be seen in those sections which were made from the *grey* part of the blood-vessel. In sections, however, which include any of the patches which look *brown* to the naked eye, it will be found that the difference of colour is due to the epithelial cells having at these parts become accidentally rubbed or washed away before the silver solution was allowed to act; for since the subjacent tissues (the sub-epithelial connective tissue, if present, and the muscular tissue of the middle coat) contain more intercellular substance than the epithelial layer (where it only occurs in fine lines between the cells), they assume a browner appearance after the reduction of the silver, and show under the microscope in the one case irregular white patches—the cell-spaces—upon the brown ground, and in the other transversely arranged lanceolate white markings—the plain muscular fibre-cells—with a variable amount of ground-substance between. The latter appearance may be obtained all over the preparation if the blood-vessel which is to be treated with silver nitrate is first brushed with a camel-hair pencil moistened with distilled water, for by this means the epithelial cells are removed. Where the sub-epithelial connective tissue is absent, the elastic layer is the only part of the internal coat which remains, and since this does not reduce the silver salt the muscular layer is the one which is seen in such cases. Except when the epithelial cells are first removed, either purposely or accidentally, even a comparatively long exposure to the action of the nitrate of silver solution will not cause the deeper coats to become stained. This is the case with all structures which are coated with epithelial cells, which appear to resist the passage of the silver salt to the subjacent tissues.

Elastic layers; fenestrated membrane; muscular tissue.— To prepare these several parts, a piece of artery (or vein) is taken (as fresh as possible, but this is not so imperative as for the preparation of the epithelial layer), and placed for two or three days in a large quantity of weak solution of bichromate of potash (about 1 in 800), or in one-third alcohol. The piece is then taken out, pinned down on a cork, with the inner surface uppermost, and a thin strip torn off from the inner surface with fine forceps. This is transferred to a slide, and teased in a drop of water. It will be found advantageous to employ only as much water as will keep the tissue moist, and to add more by placing a drop on the cover-glass before it is laid on. If the small pieces are examined, it will be found that they are for the most part made up of a close network of elastic fibres of varying degrees of fineness. Many of them have very broad fibres and small meshes, so that there may be found in different arteries every transition to the true elastic fenestrated membrane. This membrane may be met with projecting at the edges of some of the fragments of tissue, or even entirely separated; the fragments are generally curled at the edges, are often striated, and nearly always exhibit rounded holes. The fenestrated membranes are more frequently met with in the inner coat of some of the smaller or medium-sized arteries (especially the basilar), than in the largest vessels (such as the aorta).

But besides the different kinds of elastic tissue there is also to be found in nearly every such preparation a number of plain muscular fibre-cells scattered about in the fluid; for, in stripping off the inner coat, shreds of the middle coat always adhere to it, and the muscular cells readily separate after maceration. But the isolated cells present in many arteries such a ragged shapeless aspect that they would hardly be known for muscular tissue. A convincing proof, however, is the addition of a drop of weak hæmateïn solution at the edge of the cover-glass. This, as it comes to each of the cells in question, almost instantaneously stains their long rod-shaped nuclei of an intense violet, whilst the body of the cell, if the logwood solution be sufficiently weak, remains uncoloured. The addition of a little glycerine at the edge of the cover-glass, and the subsequent cementing, are sufficient to preserve the preparation. These preparations may also be stained with picrocarmine.

The connective tissue and elastic fibres of the outer coat (as well as the muscular tissue of the same coat in certain veins) can also be well seen in a teased-out preparation.

Study of the structure of the blood-vessels by means of sections.—To form a correct idea of the relative thickness of the several coats, as well as to observe the differences in arrangement in different arteries and veins, it is necessary to study them in vertical sections, *i.e.* sections made in a direction at right angles to their surface. Such sections may be either transverse or longitudinal; it will be better perhaps to choose the former direction, for the middle coat is thereby better displayed. But the vessel to be cut must first be hardened. This may be effected speedily by immersing it in spirit for a day or two, and indeed this method can be employed for nearly all the tissues and organs. It is preferable, however, in many instances to effect the process more slowly, by means of some watery fluid, such as a 3 per cent. solution of bichromate of potash or a 0·5 per cent. solution of chromic acid, since in this way the parts shrink less and consequently retain their form better; the process should, nevertheless, always be completed by means of spirit. In the case of the blood-vessels an immersion for a fortnight or three weeks in a 3 per cent. solution of bichromate of potash answers well; the pieces are then placed for a day in 50 per cent. spirit, and finally transferred to strong methylated alcohol. Here they may remain without detriment until it is convenient to prepare sections of them.

The smaller arteries and veins and the capillaries.—The elongated epithelial cells which form the walls of the capillaries, and which line the arteries and veins, are readily shown in silvered preparations of any vascular tissue. It is most convenient to choose a vascular *membrane* for the purpose, because more readily displayed, and of vascular membranes the mesentery of the frog or toad is perhaps as easy to prepare as another. The following is the mode of proceeding :—The animal (a male) having been decapitated and the spinal cord destroyed, the trunk is suspended for a few minutes by the lower limbs in order to allow the blood to drain from the body as completely as possible. The frog is then placed on its back and the abdomen freely opened. A loop of intestine is seized with forceps, drawn out and gently raised, while with a soft camel-hair pencil, moistened with distilled water, the operator carefully brushes the mesentery on both surfaces, carrying the brush in every case *from* the intestine, not towards it. This brushing serves two purposes—in the first place, the epithelial cells of the serous membrane, which would obstruct the passage of the silver solution to the blood-vessels, are removed ; and in the second place, much of the blood which remains in the vessels is driven out of them.

The brushing being completed, the loop of intestine, with its included mesentery, is cut off, rinsed in a capsule of distilled water, and at once placed in 1 per cent. nitrate of silver solution. Here it is allowed to remain ten minutes ; after which it is rinsed in distilled water and then placed in common water in the sunlight. If the day is bright, the silver is soon reduced, and all that remains to be done is to place the preparation in a shallow glass dish, and carefully cut off and remove the piece of intestine, leaving the mesentery. This must be done under water, and will require sharp scissors and delicate handling, so as not to drag upon the mesentery or throw it into folds.

In order to mount it a slide is held in the water and the membrane allowed to float over, after which the slide is carefully lifted out with the membrane flat upon it, and the excess of water is drained off. Before covering it the preparation must be examined, both with the unassisted eye and with a low power of the microscope, so as to detect any folds or creases in it. If present they can be got rid of by gently drawing out first this corner and then that with a needle. A drop of strong glycerine may now be placed on the middle of the preparation, and the cover-glass laid on and allowed slowly to settle down. More glycerine may be afterwards added at the edge if necessary. Or, in place of mounting in glycerine, the specimen may be allowed to dry on the slide, and may then be mounted in xylol balsam.

If there is but little sunlight the reduction of the silver may often be better effected by placing the loop of intestine, with its attached mesentery, after it has been taken from the silver solution and rinsed in water, in a beaker of weak spirit (equal parts of water and strong spirit, freshly prepared), and exposing it to the light in this for an hour or more. The cutting off of the intestine must be performed in the same fluid, and the mesentery floated from it on to the slide. The method has the advantage not only of effecting the reduction of the metal with greater surety, but also of rendering it more easy to obtain the membrane free from creases, for the mesentery is partly hardened by the spirit while in a state of extension, and continues in this condition when floated on to the slide, so that it is seldom necessary further to extend it by artificial stretching. In these silvered preparations little but the epithelial cells can be made out, for the rest of the tissue generally remains almost unstained, and becomes very transparent in glycerine or balsam.

To exhibit the muscular structure of the small arteries and veins, and the nuclei of their epithelial lining and of the walls of the capillaries, the vessels are stained with logwood. This is done by immersing the mesentery or other vascular membrane, after having lain for a few hours in one-third alcohol or 1 in 800 bichromate of potash solution, in a dilute solution of hæmateïn, until distinctly coloured ; then place the tissue in water, and mount it, with the same precaution as before to prevent creasing, in glycerine. For the structure of small arteries the pia mater from the human brain may be used. A small piece is stripped off with forceps ; and as it consists almost entirely of small arteries and veins, and moreover a few capillaries are generally dragged out with it from the cerebral substance, the structure of all these vessels is, after staining, very well displayed. The small veins are here exceptional in being entirely devoid of a muscular coat, whereas the arteries have this coat well developed, and it is particularly well shown in consequence of the staining of the transverse nuclei of the muscular fibres by the logwood. Within these may be detected, by carefully using the fine adjustment of the microscope, in the first place longitudinal striæ, which are produced by wrinklings in the elastic layer of the internal coat ; and in the second place, situated most internally, elongated oval nuclei, which belong to the epithelial cells lining the vessel. Of course two layers of all these structures are come across in focussing from above down. The outer coat is represented merely by a few corpuscles and fibres of connective tissue which blend externally with the connective tissue framework of the membrane.

It will be found, in carefully focussing from above down, that at one position of the focus the small vessel has exactly an appearance as if it had been cut longitudinally through the middle, and as if the top of the lower half were being examined. The lumen is seen in the centre, with possibly a few blood-corpuscles still in it ; on either side of this a well-marked line representing the inner coat ; outside this again what seems like a row of rounded cells, which are really the encircling fibre-cells of the muscular coat seen as if cut across ; and finally, here and there, outside of all, small cells presenting the fusiform aspect of connective tissue corpuscles seen in profile. All these appearances are exactly the same as if a section had been made along the vessel, and result from the fact that only those parts of an object which lie in the horizontal plane that happens to coincide with the focal distance of the objective are distinctly seen, so that it seems as if only this particular slice were present. An 'optical longitudinal section' is thus obtained of the vessel.

STUDY OF THE CIRCULATION

The study of the blood-vessels cannot be said to be in any sense of the word complete until they have been viewed in the living condition and with the blood still moving through them. Such an observation can, of course, only be made whilst an animal is still alive, and in parts which are transparent enough to allow the vessels to be distinctly seen. Membranous parts are those which are naturally best adapted for such observation, as, for example, the web of the frog's foot and of the bat's wing, the tails of tadpoles or of small fishes; the tongue, mesentery, and lungs of the frog and toad, but especially the latter animal; and the mesentery and omentum of small mammals. In such preparations the surrounding tissues, and especially the connective tissue corpuscles, may be studied as well as the blood-vessels; and the changes due to commencing inflammation which are exhibited by the bloodvessels, and the migration from the veins of the white bloodcorpuscles, can always be brought on either by the application of irritants or, as in the case of the serous membranes, by simple exposure to the air. The best methods, therefore, of observing the circulation of the blood in different parts will be described in the following preparations.

Circulation in the frog's web.—One of the common English frogs (*Rana temporaria*), of as light a colour as possible, is placed in water with which chloroform has been shaken up. In this it soon becomes immobilised, and the anæsthesia may be easily kept up for hours by keeping the skin moist with the chloroform water. Or a few drops of a very weak solution of curare (1 to 1,000 of water) are injected under the skin of the back. This is generally sufficient, in the course of from a quarter to half an hour, to render the animal completely motionless, whilst the pulsations of the heart and the circulation proceed unimpaired. The frog is then laid on a plate of glass of an oblong shape and one of its legs is stretched out with the webs resting on the glass. If necessary, the web under examination may be slightly stretched by being fastened on either side by means of threads attached to the adjoining glass by modelling wax. Care must be taken that the web is not tightly stretched, since this tends to arrest or obstruct the circulation. A slip of blotting-paper or a piece of linen rag is placed over the animal and kept wetted with water, and the glass, with the frog upon it, is then placed on the stage of the microscope (the head of the animal being away from the observer, and the web over the aperture in the diaphragm), and is fixed in this position by the clamps, like an ordinary slide. To observe the web a low power is used to see the general features of the circulation, a high power being afterwards employed to observe the parts more in detail. But this should not (unless it is an immersion) be of too short a focal distance, since otherwise the lower glass is apt to become clouded by moisture from the web. It is not advisable to put a piece of covering-glass on the latter to prevent the clouding, as the circulation is thereby liable to be interfered with or arrested. A cork frog-board with a slit at one end (fig. 51, *a*) may be used instead of a glass plate, and the web gently stretched over the slit by pin points.

The web of the frog's foot is the easiest of the vascular membranes to prepare, but has the disadvantage that, owing to the thickness of its epidermic coverings, it is not always easy clearly to make out the intermediate tissues. At the same time, being under almost completely natural conditions, the circulation will continue for an indefinite time quite unimpaired.

Circulation in the tails of tadpoles and fishes.—The mode of studying the circulation in these requires no special description.

A tadpole can be readily immobilised by placing it in a putty cell, through a hole in one side of which the tail is allowed to protrude. For small fishes a special box devised by Caton is used. The fish is fixed securely in this with the tail projecting, and a stream of water is allowed to flow over the head and gills.

Circulation in the mesentery.—The mesentery of the frog, and still better of the toad, is admirably adapted by its thinness and perfect transparency, as well as its vascularity, for observations on the blood-vessels and surrounding tissues. It

Fig. 51

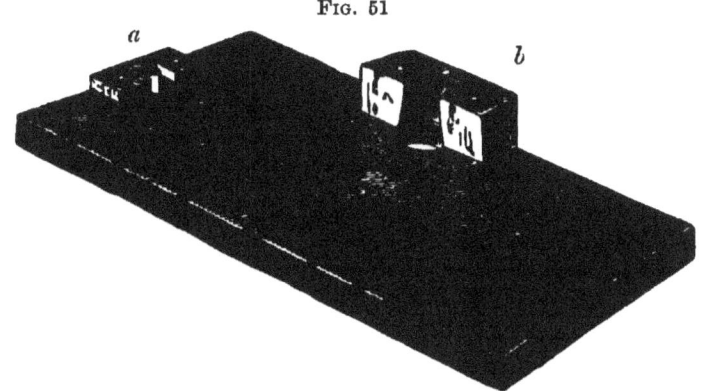

Flat piece of cork arranged as a frog-stage for viewing the circulation in the web, tongue, mesentery, or lungs. About half natural size

Over the small piece of cork *a* the tongue can be fixed ; *a* can be removed when the slit below it is wanted for the web ; *b*, cork with a deep groove cut along one side : to this the intestine is fastened by needle points, while the mesentery rests on a semicircular piece of glass which should fit at the top of the groove

is necessary to have a special mesentery-plate for this purpose, which can, however, be readily made from an oblong piece of cork or soft wood. A round hole about half or three-quarters of an inch in diameter is made at one side with a cork-borer, and a small piece of cork about half an inch thick, and with a segment of a similar circle cut out of its side, is fixed on the board with sealing-wax or small pins (fig. 51, *b*). A piece of glass of the same shape as this segment may be fitted into it near the top for the mesentery to rest on. The animal—preferably a male—having been rendered insensible by destruction of the brain, is curarised as before, and laid upon its back, and a longitudinal cut about an inch long is made with scissors through the skin of the abdomen, about half an inch to the right of the middle line. Before proceeding further, the operator should wait for a minute or two to make sure that there will be no bleeding; and any blood that may have already exuded should be dried up with blotting-paper. The abdominal cavity is then opened by a corresponding cut through the muscles and peritoneum, taking care to avoid, if possible, any veins that may be seen. Having again assured himself of the absence of bleeding, the operator very gently draws out one of the coils of the intestine, with its included mesentery, at the aperture. The animal is now to be turned over on its side, and so propped up against the smaller cork that the wound is about on a level with the top. All that remains to be done is to place the extruded mesentery over the aperture, and, if necessary to keep it in position, to pass two or three fine needle-points through the intestine into the cork. In this case, again, the greatest care must be taken in no way to drag upon the exposed membrane, nor to allow it to be pressed upon. Moreover, the surface must from time to time be moistened with salt solution, to prevent its becoming dry. But it will be found that the mere exposure of the serous surface to the air is sufficient to produce before long the changes in the circulation (dilatation of vessels, sticking and migration of leucocytes) which are characteristic of the commencement of inflammation.

Capillary circulation in mammals.—It is less easy to study the circulation in the serous membranes of mammals, for the exposure required for the purpose is apt to be far more prejudicial to the maintenance of the normal condition of the tissues than is the case with the cold-blooded vertebrates. It is necessary, moreover, to maintain the exposed part at the body temperature, and to immerse it in fluid, since it would otherwise become at that temperature rapidly desiccated. The membrane generally chosen is not the mesentery but the omentum, which in many animals, *e.g.* the guinea-pig, is very extensive, and at the same time thin, and provided in parts with a sufficient number of blood-vessels. The animal, which should be rather a small one, is anæsthetised with chloral hydrate, one or two cubic centimetres of a 20 per cent. solution being injected under the skin. The warm stage (fig. 37) is in the meanwhile got ready, and a glass tray (which can be extemporised out of a small plate of glass, some pieces of glass rod and sealing-wax) is placed on it and filled with 0·9 per cent. salt solution, which is maintained at about 38° C. Then the animal is supported on a block at a convenient level, and the abdomen having been carefully opened, a little of the omentum is drawn out and allowed to float flat in the warm salt solution, where it can be examined either with a low power or with an immersion objective dipping into the solution. If the latter is employed, a piece of thin covering-glass must be placed over the part of the membrane which is to be examined, so as to sink it in the fluid and keep it steady. But in spite of every precaution, the circulation under these conditions retains its normal character but a short while, and inflammatory congestion and *stasis*, or complete stoppage of the flow of blood, rapidly supervenes.[1]

Circulation in the lung of the toad.—This is readily observed with the aid of the mesentery board. The animal must, as before, be first rendered insensible and curarised : it will be found that a good-sized toad will require at least six times as much curare as a frog. An opening is made at the side of the chest large enough to allow the lung, which in the curarised toad almost always remains distended with air, to protrude. The animal is then propped up on the mesentery board (fig. 51) in such a manner that the lung rests over the aperture *b*, and the circulation can be studied in the part which is uppermost without further trouble. A frog may be used in a similar way, but there is much greater difficulty in keeping the lung distended. This can, however, be done by the use of a special apparatus devised by Holmgren. In either case the greatest care must be taken to avoid pricking, or in any way rupturing the wall of the lung.

[1] For a detailed account of this method the student is referred to the original description by Burdon Sanderson and Stricker in the 'Quarterly Microscopical Journal' for 1870.

174 PRACTICAL HISTOLOGY

Circulation in the tongue of the toad.—By far the most beautiful object for studying not only the circulation but also the tissues in the living animal, is the tongue of the toad, and in a slightly less degree that of the frog. The tongue is in these creatures an extremely extensile organ, which, under

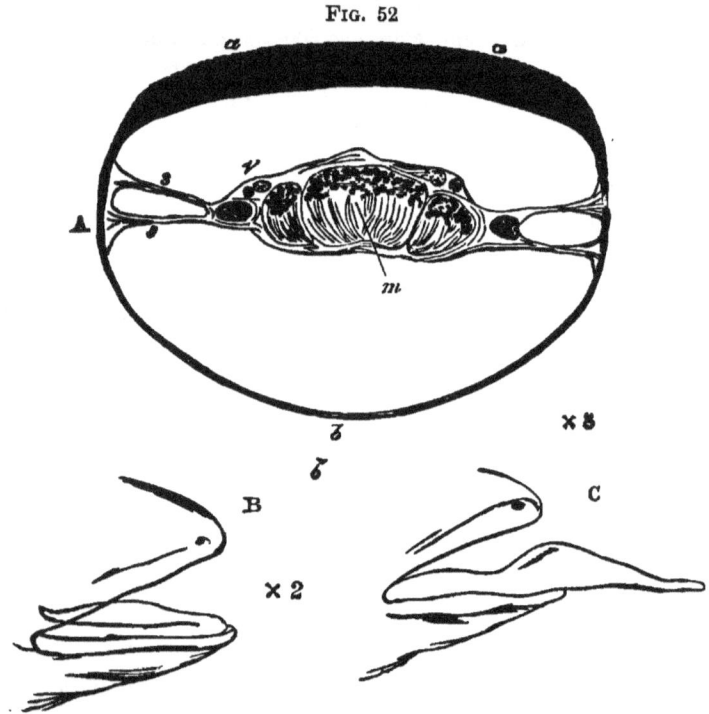

FIG. 52

Structure and position of the tongue of the toad (Dowdeswell)

A. Transverse section through the middle of the organ with the lymph-spaces fully distended; *a a*, thick, papillated mucous membrane; *b*, thin lower membrane; *m*, muscular bundle cut across, united to the sides of the tongue by septa of connective tissue, *s s*; *v*, position of the larger blood-vessels
B. Profile view showing the tongue in its ordinary position within the mouth
C. The same when extended

ordinary circumstances, lies folded back on the floor of the mouth (fig. 52, B), but which can at the will of the animal be protruded for a considerable distance (c). For the preparation of the organ the frog-board is again necessary, a smaller piece of cork of the shape shown in the figure (fig. 51, *a*) and about one-eighth of an inch thick, being fastened with pins over the slit which served for the display of the frog's web; to its aperture a thin piece of glass may be adapted.

The toad having been rendered insensible and curarised as before, is laid upon its back with its nose close to the slit. The lower jaw is then raised and the folded back end of the tongue is found, and drawn gently out of the mouth with forceps. The end has a pointed projection or cornu on each side; these are successively laid hold of by the forceps and fastened with needle-points to the small piece of cork on either side of the slit. Before the rest of the operation is described, a word or two may be said with regard to the structure of the organ. It is not solid throughout as in mammals, but hollow, the interior being occupied by a lymphatic cavity. This lymph-space is traversed by bundles of muscular fibres (fig. 52, m) which pass towards the extremity of the organ and are connected to the sides by delicate septa of connective tissue ($s\ s$). Above the lymph-space in the present position of the animal—on its back with the tongue extruded—the mucous membrane is thick and papillated (fig. 52, A, $a\ a$). Below is a very thin and delicate mucous membrane (b), with numerous blood-vessels, and small muscular fibres running over it. The former membrane is too thick and irregular to allow the delicate internal structures to be seen through it; it is therefore slit up longitudinally and pinned to either side. But to do this without injuring the parts below, it must be well separated from them, and this can best be effected by distending the lymph-sac with salt solution. With this object a hypodermic syringe, provided with a fine and sharp cannula, is filled with the fluid, and its point is stuck into the tongue near the end, passing about half an inch backwards. It will almost certainly be found that on pressing the piston down the salt solution will readily flow into the lymph-sac and, as it fills this, will cause the thin mucous membrane at the lower part to become bagged out and completely separated from the muscular bundles, *m m*, and these again from the thick layer above. The latter is now carefully slit up along its middle by sharp fine scissors, and first one edge of the wound and then the other is drawn to the side of the slit in the cork and fastened there by two or three needle-points. If everything is carefully done there will be no escape of blood over the preparation; but should any blood have exuded it may be washed off by pouring a little salt solution over the surface.

There is now brought to view the fan-like group of muscles which pass through the middle of the lymph-sac, and the bundles of which are, as before mentioned, connected with the sides and with one another by delicate septa of connective tissue, traversed by a few blood-vessels; and it is this delicate connective tissue, of which two strata can generally be traced, one superficial to the other, which is best adapted for exhibiting both vessels and connective tissue corpuscles. Moreover, the mere exposure of the lymphatic surface soon causes inflammatory changes, and after the preparation has been made a few minutes only, the commencement of these is seen in the sticking of the pale corpuscles to the walls of the vessels, speedily followed by their migration from the veins into the surrounding tissue. Nowhere can the fact be more clearly established, and the details of the process more accurately followed than here. The circulation of the blood among the muscular fibres can also be well seen in this part of the tongue.

Lastly, by focussing through the connective tissue septa, or by severing the longitudinal muscular bundles which they serve to unite, the vessels of the lower mucous membrane are brought into view, especially if a slip of glass is fitted into the small piece of cork, so as to support the tongue and prevent the thin membrane from bulging downwards.

METHOD OF INJECTING THE BLOOD-VESSELS

Before leaving the subject of the blood-vessels the best mode of filling them with transparent material may be described, especially as in the study of the several organs it is necessary, in order that the course and arrangement of the vessels may be properly made out, that sections of injected as well as of uninjected preparations should be looked at. It will be convenient in this place to describe the injection of a small animal entire from the aorta, reserving any special directions concerning organs which are not thereby properly injected, such, for instance, as the lungs and liver, until they are severally dealt with.

Preparation of the injection mass.—This is almost always a solution of gelatine coloured either red with finely precipitated carmine, or blue with soluble Prussian blue. Sometimes, but rarely, when it is wished to inject two sets of vessels of different colours, both of these are used, but as a rule all the blood-vessels—arteries, capillaries, and veins—are filled with the same injecting fluid; preparations in which the arteries are filled with one colour and the veins with another are difficult to prepare, and present no practical advantage. The gelatine solution is made as follows :—Ten grammes of clear gelatine cut into small pieces is placed in a beaker containing 50 c.c. of cold distilled water. In an hour or two the gelatine will have swollen to several times its original volume. A glass cover is put over the beaker, and it is placed in a water-bath and heated until the gelatine is rendered fluid.

For the *red* injection 4 grammes of carminate of ammonia is rubbed up in a mortar with 50 cubic centimetres of distilled water. When the carminate of ammonia is as completely as possible dissolved the liquid is filtered and the filtrate is warmed. The gelatine solution is then gradually added to it with constant stirring. The next part of the process is to precipitate the carmine, for otherwise it would diffuse through the walls of the vessels and colour the tissues; but it must be precipitated so finely that the particles shall not be visible even under the highest power of the microscope. To effect this object the solution is to be acidulated by running a small quantity of 10 per cent. solution of acetic acid from a burette drop by drop into the warm carminised gelatine solution, which is all the while constantly agitated. The alteration in reaction may be shown, in spite of the red colour of the solution, by placing a small drop on the *back* of a piece of glazed blue litmus paper ; if the coloured face of the paper be looked at it will be found to have assumed the characteristic red which acids produce, and which is quite different from carmine. This change is caused by the diffusion of the acetic acid through the paper, whereas the carminised gelatine sets almost immediately, and is thus unable to soak through. Or a drop of the injection-mass to be tested may be put on the glazed coloured face of the litmus paper, and after half a minute wiped off with a piece of linen wetted with distilled water. The spot where it had rested will, if it be acid, show characteristic change in colour.

It is not sufficient that the fluid should only just be acidulated ; there must be an excess of acid. A few more drops are therefore added, and the carmine completely thrown out of solution. This change from the soluble to the insoluble state is accompanied by a marked alteration in colour, for whereas whilst still in solution the carmine imparted the rich deep red of an ammoniacal solution to the gelatine, after the precipitation the colour of the latter changes to a paler red, comparable rather to the tint presented by the powdered carmine in the dry state. Even after the production of this change a few more drops of the acetic acid may be added, for it will do no harm, and will tend to counteract the natural alkalinity of the tissues.

The coloured gelatine is next strained through a piece of flannel or fine linen, previously soaked in hot water and again wrung out, and is collected in a flask as it runs through the filter, and transferred to the injecting-bottle.

For the *blue* injection 10 grammes of gelatine is taken, and after having been soaked in cold water and dissolved up as before, 50 c.c. of a 2 per cent. solution of Berlin blue, which has been previously warmed, is gradually added with constant agitation to the fluid gelatine. The blue mixture is strained, and is then ready for use, without the necessity of precipitating the colouring matter, for this being colloid does not diffuse readily.

It is sometimes advantageous in cases where the structure of the walls of the blood-vessels is to be the subject of observation, to use an injection-mass which is far less deeply coloured. This can of course be readily obtained by diminishing the proportion of carmine or Berlin blue which is used.

The soluble Berlin blue is of great value for the purpose of injecting both the blood-vessels and lymphatics. It is somewhat troublesome to prepare. The following is the method recommended by Brücke, to whom we owe its introduction:—

Take of potassic ferrocyanide 217 grammes, and dissolve in a litre of water (solution A).

Take a litre of a 10 per cent. solution of ferric chloride (solution B).

Take four litres of a saturated solution of sulphate of soda (solution C).

Add A and B each to two litres of C. Then with constant stirring pour the ferric chloride mixture into the ferrocyanide. Collect the precipitate upon a flannel strainer, returning any blue fluid which at first escapes through the pores of the flannel; allow the solutions to drain off; pour a little distilled water very carefully over the blue mass, returning the first washings if coloured, and renew the water from day to day until it drips through permanently of a deep blue colour. This is a sign that the salts are washed away, and all that is further necessary is to collect the pasty mass from the strainer and allow it to dry.

Apparatus employed for injecting.—This consists, in the first place, of a bottle for holding the coloured fluid; and secondly, of some means of producing a steady, elastic, and readily alterable pressure on the surface of the fluid so that it may be driven with any required force into the arteries. Fig. 53 represents a convenient form of apparatus for general use. The bottle (i),[1] which holds the injecting fluid, is a moderate-

---

[1] A similar one is shown on a larger scale in fig. 56, at c.

sized, wide-mouthed phial, with a well-fitting vulcanised indiarubber cork, through which three glass tubes pass.

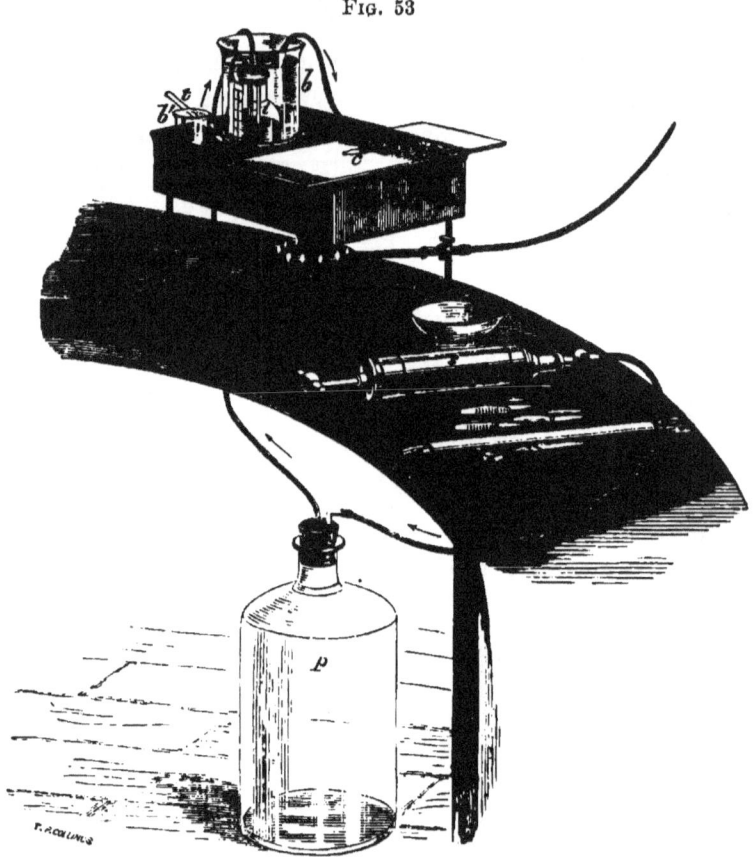

Fig. 53

Injecting apparatus. Complete

s, condensing syringe, fixed to the table; *p*, pressure bottle; *b*, beaker of warm water in which the injection-bottle, *i*, stands; *b'*, small beaker containing salt solution; *w*, water-bath heated by a ring burner below : the temperature of the water is indicated by a thermometer, *t*, placed in it ; *c*, arterial cannula, connected to an indiarubber tube from the injecting-bottle ; close to the cannula is a steel-clip. The cannula rests upon a glass plate, which may serve either to put the animal which is being injected upon, or to cover it over, if it is thought necessary to place it in the water-bath

One of these goes to the bottom, and from it an indiarubber tube passes, which will be subsequently connected with the artery cannula (*c*), but not before this has been inserted in the blood-vessel, in the manner immediately to be described. The other passes only just through the cork, and serves to maintain communication by means of another indiarubber tube with the pressure-bottle p. The third tube is connected with a pressure-gauge, which may be a small mercurial manometer. The injection-bottle is placed during the process of injecting in a glass vessel (b) of warm water (about 40° C.) ; a piece of cork is wedged in between the bottle and the side of the vessel to prevent the bottle from floating up as it becomes emptied of injection, and the vessel is covered with a glass plate (not shown in the figure). The pressure-bottle is a large glass or earthenware bottle of from one to three gallons capacity, and tightly fitted with an india-rubber cork, through which two glass tubes pass. One of these is connected, as before mentioned, with the injection-bottle, and the other with a condensing syringe (s), by means of which the air within the bottle can be brought to any state of tension that may be desired. Finally, if the injection is to occupy a considerable time, a water-bath (w), heated by a ring-burner to about 40° C., should be provided for receiving both the beaker containing the injection-bottle and the animal, and maintaining their temperature during the process. Ordinarily, however, if the operation be quickly and dexterously performed, the whole process will not occupy more than a few minutes, and will be over before the natural heat of the body has had time to become dissipated.

Everything then being in readiness, the animal—a rabbit, guinea-pig, or rat, for example—is killed by chloroform inhalation, being placed under a bell-glass with a sponge wetted with chloroform. As soon as it is dead it is taken out and held by an assistant, whilst the operator first quickly reflects the skin from the front of the thorax and then makes an opening in that cavity just over the position of the heart. This is then seized near the apex with blunt forceps, drawn out of the aperture, and held here by an assistant. The aorta is then found, the point of a pair of forceps passed under it close to the heart, and a thread ligature drawn round it. A snip is now made in the left ventricle, and an arterial cannula (fig. 54 $c.^1$, $c.^2$, $c.^3$) passed through this into the aorta, in which it is tied by the ligature. Then by means of a pipette a little warm water or salt solution is passed into the cannula so as completely to fill it to the exclusion of air.

The next thing to do is to connect the cannula with the

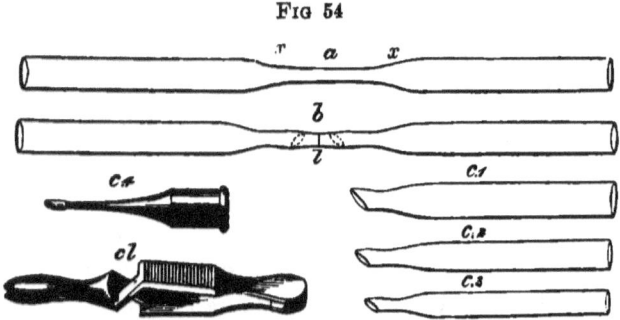

Cannulas for injecting. Natural size

$c.^1$, $c.^2$, $c.^3$, glass cannulas of different sizes; $c.^4$, metal cannula; it is sometimes more easy to insert than the glass ones, especially into fine blood-vessels, or into lymphatics. cl, steel clip for clamping an artery, or a small indiarubber tube. a and b are intended to illustrate the mode of making the glass cannulas; a, glass tube which has been heated in the middle in the blow-pipe flame, and drawn out so as to be narrower here; b, the same tube after having been again heated (by the tip of the flame), and drawn out at the points $x\ x$, so as to narrow it still more at those places. The subsequent proceeding consists in making a nick at l with the edge of a file, breaking the tube across here, and with a fine, flat, wetted file grinding the ends away obliquely as far as the dotted ring in each. The sharpness of the filed edge is got rid of by inserting it for a moment or two in the flame. Two similar cannulas are thus made from one piece of tubing

indiarubber tube which brings the injection from the bottle. But this tube must first be completely filled by the injection, so that it contains no bubble of air. To effect this, whilst the indiarubber tube is kept closed by the spring clip with which it is provided, the air in the apparatus is put under a pressure of about two inches of mercury by working the syringe. The free end of the indiarubber tube is now held up, and the clip opened until the coloured fluid forced up by the pressure begins to escape, when the clip is immediately closed, and the tube is slipped on to the arterial cannula. The greatest care must be taken throughout to avoid the introduction of air, since this would obstruct the smaller vessels and prove fatal to the success of the injection.

The clip is now permanently opened and the injection suffered to flow into the aorta, at first under the low pressure of two inches of mercury ; but the pressure is gradually increased by working the syringe until a pressure of three or four inches is attained. The blood in the vessels gets forced before the injection into the right cavities of the heart, so that these are soon much distended ; when this is the case the right ventricle is slit open and the accumulated blood allowed to flow out. The blood is soon followed at first by a mixture of blood and the coloured gelatine, but afterwards by the latter only ; after this has been escaping for a minute or two, the slit in the ventricle is closed by placing a clip on, or tying a tape round the heart, and the injection, being thus obstructed in its outflow, accumulates in the vascular system and distends all the vessels to their fullest extent. The success of the injection may be estimated by the colour of those parts which are not concealed by the fur—the paws, lips, nose, and ears, for instance—and the tongue and interior of the mouth. After waiting a few minutes longer, until the vessels may be considered to be completely filled, a tape ligature is put round the base of the heart, so as to include all the great blood-vessels, and is slowly tightened. This will effectually prevent any escape of the fluid gelatine from the vessels when the cannula is removed from the aorta. This may therefore now be effected, the clip having been first replaced on the indiarubber tube which is connected with it, so as to prevent the injection from spurting out.

The animal is then put aside until it has become cold and the gelatine is firmly set. Any parts which are wanted are then cut out and placed either in 60 per cent. spirit, to be afterwards gradually transferred to strong spirit, or they may with advantage be placed in 3 per cent. solution of bichromate of potash, in which they remain for a fortnight or longer, or in 10 per cent. formol.

The blue injection possesses the disadvantage that in spirit it is apt to become temporarily reduced in the smaller vessels and rendered almost colourless, so that it is difficult to determine whether a successful injection has been made or not. The colour may, however, be readily restored by pouring some oxidising fluid, such as spirits of turpentine or a weak solution of peroxide of hydrogen, over any part about which there is doubt; and in the ordinary course of preparing and mounting sections the blue colour is always brought back, especially if turpentine is used as a clearing fluid.

The preparations in bichromate may conveniently be cut by the freezing method.

Most other forms of apparatus which are used for injecting are more or less like the one above described, the chief modification being in the mode in which the pressure is produced, this being effected in one form by allowing water to flow from a tap into the pressure-bottle (which in such cases is generally made of metal), and thus compressing the air; in another by allowing mercury to flow from one vessel into another. But the latter method, although useful when small quantities of injecting-fluid only are required, as with the injection of the lymphatics (with which the apparatus will be described), is costly for large quantities; and it will be found that none are more simple and efficient in working than the one here recommended.

If a condensing syringe is not at hand, sufficient pressure may be got in many cases merely by *blowing* air into a pressure-bottle through an indiarubber tube, its escape being prevented by subsequently clipping the tube.

Injections which are fluid in the cold (of which the best is a 1 or 2 per cent. solution of Berlin blue) are sometimes used for the blood-vessels, especially for injecting cold-blooded animals.

Any of the coloured gelatine that may remain over can be

preserved until again wanted, if the precaution is taken, after disconnecting from the pressure-bottle and allowing the fluid in the cannula tube to run back, to place the bottle, tubes and all, for twenty minutes in boiling water, and whilst still hot to stopper up the ends of the tubes with pieces of glass rod. The whole can then be put away until wanted.

If it is desired to show the endothelium of the small blood-vessels the silver-gelatine mass which is recommended on p. 210 for showing the epithelium of the pulmonary alveoli may be employed, or simply a solution of silver nitrate of 1 per 1,000 in distilled water may be used. But in either case the vessels should first be washed out with a 2 per cent. solution of nitrate of soda in distilled water.

CHAPTER IX

LYMPHATICS. SYNOVIAL MEMBRANES. SEROUS MEMBRANES

IT is in preparations of the serous membranes that the structure and arrangement of the lymphatic vessels can be best demonstrated, and it will on this account be convenient to combine them here under one head, especially as the method which on the whole exhibits the structure of the serous membranes best is the only one which shows at all satisfactorily the structure of the lymphatic vessels and their relation to the cell-spaces of the connective tissue.

Preparation of the rabbit's omentum by the silver method. A rabbit (half-grown) having been killed by bleeding, the abdomen is opened, and the omentum, which is generally to be found crumpled up close beneath or to the left of the stomach, is raised with forceps, cut off as close to the line of attachment to the stomach as possible, and placed in a shallow dish of 2 per cent. solution of nitrate of soda in distilled water which is at hand to receive it. Besides this solution there should be ready on the table a little 1 per cent. solution of bichromate of potash, some one-half per cent. solution of nitrate of silver, a wash-bottle of distilled water, a flat glass dish containing a mixture of spirit and water (equal parts), two glass plates about four inches by six, a large soft camel-hair brush, and two or three clean capsules and watch-glasses.

Two small corners are first to be cut off the omentum. One of these is placed in the bichromate of potash, put aside, and examined after two or three days; this is for exhibiting the arrangement of the connective tissue fibres. The other is first rinsed with nitrate of soda solution, then placed for one minute in a watch-glass containing a little of the silver solution, rinsed again with distilled water, and exposed to the sunlight in another watch-glass containing water. After from a few minutes to half an hour of exposure, according to the intensity of the light, it may be removed, and a portion or the whole of it cautiously mounted by being floated upon a slide under water. The excess of water is removed from the slide, all creases and folds are carefully got rid of in the same way as with the frog's mesentery, before described, and finally either a drop of glycerine is added and the cover-glass superposed, or the preparation is dried and mounted in xylol balsam. This preparation is for the purpose of showing the epithelial layer which covers each surface of the membrane. If only a portion were mounted, the rest may be placed for a few minutes in hæmalum before mounting; in this way the nuclei of the cells may be brought to view.

But while this second corner was being exposed to the light the preparation of the rest of the omentum can be proceeded with.

In the first place, it is floated on to one of the glass plates and removed from the fluid, and then by drawing gently first at one place and then at another the creases and folds are gradually removed, and it is in this way spread out as an exquisitely delicate membrane, which may be made to cover the whole upper surface of the glass plate, and may be extended round its edges so as to reach the lower surface. As soon as all folds are in this way removed from the part which covers the upper surface of the plate, the second glass plate is applied to the under surface of the first, and the membrane, or at any rate its greater part, in thus maintained in an extended state. The two glass plates can then be fixed together by indiarubber bands. Next the surface is gently brushed all over with the camel-hair pencil moistened with nitrate of soda solution; this is for the purpose of removing the epithelial layer from that surface, and enabling the silver solution more rapidly to penetrate. The brushing is not absolutely essential, for in many parts, especially those in which the lymphatics and blood-vessels are most numerous, the epithelial layer is deficient, or at least incomplete and modified.

The solution is now drained off the membrane as much as possible, and without delay a quantity of nitrate of silver solution is poured on it and allowed to run over every part of the exposed surface. After five minutes it is washed away with distilled water, and the glass plates, with the membrane of course still upon the surface of the upper one, are placed in 50 per cent. spirit contained in a flat glass dish. The dish is then covered and placed in the sunlight until the silver, as evidenced by the change in colour, is fully reduced. Small pieces, each about an inch square, are then cut with sharp scissors from various parts, floated on to slides with the browned surface uppermost, and are exposed to the air for a few minutes to allow most of the *spirit* to evaporate, leaving them in water; to this a drop of glycerine is added, and finally the cover-glass is superposed. Or the preparation is allowed to dry on to the slide and is then simply mounted in xylol balsam. Before mounting in glycerine, the cell-nuclei may be stained by hæmatoxylin after the silver.

Both before and after the treatment with nitrate of silver it may have been noticed that the delicate membrane is studded all over with patches of thicker tissue, some quite small and insular, others extending over a considerable area. These patches, which are characterised by an accumulation of lymphoid cells and by the small size of the epithelial cells of the surface, are—at least the larger ones—provided with numerous blood-vessels, the epithelium of which is often very well stained by the silver; and these are always accompanied by one or more lymphatic vessels, with walls formed by the characteristic wavy outlined cells.

The mesentery.—A piece of the mesentery can be prepared in a manner similar to that employed for the omentum. But the glass plates used for stretching the membrane must be smaller, and they may be advantageously replaced by Hoggan's rings. These consist of a broad and slightly conical ring of vulcanite, an inch or more in diameter, over the smaller end of which a narrow ring slips. This smaller end is applied to the piece of membrane which it is desired to use, and the narrow ring being then slipped over, the membrane is securely fixed and stretched over the larger ring, which it converts into a cup, the bottom of which is formed by the membrane. The remainder of the tissue is cut away. Each surface of the membrane can be readily brushed, and the silver nitrate solu-

tion can be poured into the cup, and thus remain in complete contact with the membrane. A wooden ointment-box with the top and bottom removed forms a rough substitute for the vulcanite rings.

The central tendon of the diaphragm with its serous coverings.—The thoracic and abdominal surfaces of the diaphragm present important differences in the arrangement of the numerous lymphatic vessels which are distributed upon them. To see both properly it will be necessary to sacrifice two animals. It is best to use rabbits, since their central tendon is larger and thinner in proportion than that of most other mammals.

For the preparation of either side both thoracic and abdominal cavities must be freely opened, the marginal attachment of the diaphragm being left intact, so that that muscle remains stretched out. A double ligature is to be put on the inferior vena cava in the thorax, and the vessel cut between the two threads ; this is to prevent the blood in the vessel from getting over the membrane. The pericardium is now cut away from the upper surface of the diaphragm, and the suspensory ligament of the liver from the lower. The side upon which it is wished to display the lymphatics is then brushed pretty firmly with a camel-hair pencil wetted with nitrate of soda solution or distilled water, after which a few drops of nitrate of silver solution are allowed to flow over it, or are applied with the brush. After five minutes' contact the silver solution is washed off by a stream of distilled water, and the central tendon, including also some of the muscular fibres which converge to it, is carefully removed, pinned out upon a loaded cork or cake of wax, with the silvered surface uppermost, and exposed to the sunlight either in water or 50 per cent. spirit. When distinctly browned it is removed from the window, and pieces from different parts are cut out and mounted in glycerine or, after drying, in xylol balsam, as in the preparation of the omentum. The preparation can also be made with the aid of Hoggan's rings.

Stomata.—In addition to these preparations—which exhibit the lymphatics and cell-spaces of the serous membranes, and, on the abdominal side, the lymphatic clefts between the tendon bundles—it is useful to make another silvered preparation, unbrushed, of the peritoneal surface, a third animal being sacrificed for the purpose. This serves to show the epithelial layer of the serous membrane, with the differences in character of its cells in different parts, the cells being much smaller over the interfascicular lymphatics than elsewhere. Amongst these smaller cells may be seen here and there the minute darkly-stained angular patches known as pseudostomata, which are probably merely accumulations of intercellular substance; and also—but these are more difficult to find—the true holes or stomata surrounded by a ring of small cells, and leading by a short canal into the lymphatic below. Such stomata or orifices leading from the serous cavities into lymphatic vessels are met with occasionally in preparations from most of the serous membranes in mammals. But they are especially numerous and well seen in the peritoneum of the frog.

A male frog should be killed for the purpose, and the intestines and stomach removed so as to expose the back of the abdomen, but without cutting the mesentery too near the spinal column. If the trunk of the animal is now placed in a dish of 2 per cent. nitrate of soda solution, and the posterior part of the peritoneum carefully examined under that fluid, it will be found that it does not closely cover the vertebral column, great vessels, and other structures which are found at the back of the abdomen, but is separated from them by a large lymph-space, divided from the serous cavity by a membrane. This is covered on the one side by the epithelial cells of the peritoneum; on the other by those of the lymphatic, and to the unassisted eye appears to form a complete septum. Under the microscope, however, it is seen to be studded by very numerous apertures, which can be seen, even in the fresh condition, if the membrane is removed and examined with a high power. But to study their structure and the arrangement of the epithelial cells with reference to them, and also to obtain a permanent preparation, the septum is to be stained with silver. With this object the whole septum, or a portion only, is dissected off under nitrate of soda solution. It is convenient to remove with it the elongated kidneys which adhere to it behind, and to cut these away only after the staining is completed. The membrane is placed in a 0·5 per cent. silver nitrate solution for one minute, again rinsed, and exposed in water to the light. After the metal is reduced the preparation is floated on to a slide, and, with the usual precautions to avoid folds and creases, finally either mounted in glycerine or dried on the slide and mounted in xylol balsam. It is desirable, before mounting, to stain the tissue with hæmateïn so as to exhibit the nuclei of the cells, but this is not absolutely necessary, and much increases the risk of producing folds in the membrane.

For studying the structure of the larger lymphatic vessels, as, for example, the thoracic duct, precisely the same methods, both for teased preparations and for sections, are employed as were used for the larger blood-vessels.

INJECTION OF LYMPHATICS

The minute lymphatics of a part may, where numerous, generally be readily displayed by simply sticking a very fine cannula into the tissue, and forcing a coloured fluid through this. The best apparatus for the purpose of obtaining the requisite pressure is the small mercury apparatus shown in the accompanying figure (fig. 56). The mercury contained in the bottle *a* compresses the air in the pressure-bottle *b*, according to the height *a* is raised above *b*, this height being regulated with the greatest nicety by the screw *d*. The bottle *c* containing the injection-fluid communicates by one tube with the pressure-bottle, and by another (which passes to the bottom) with the injecting-cannula *f*. Gelatine is not used for injecting the lymphatics, but almost always injections which are fluid in the cold. Berlin blue solution (2 per cent.) is often employed, but the best fluid for the purpose is a solution either of alkanet or of asphalt in turpentine, either of which readily flows into the lymphatics. The cannula can be made from a piece of glass tube drawn out to a capillary point, but the best are long perforated steel needles like those supplied with hypodermic syringes, and as fine as it is possible to procure them (fig. 57). The india-rubber tube connected with the cannula is closed by the clip *g*.

The mode of injecting the lymphatics of a tendon may be here described as an example, especially as the subject was deferred when studying the minute structure of tendon.

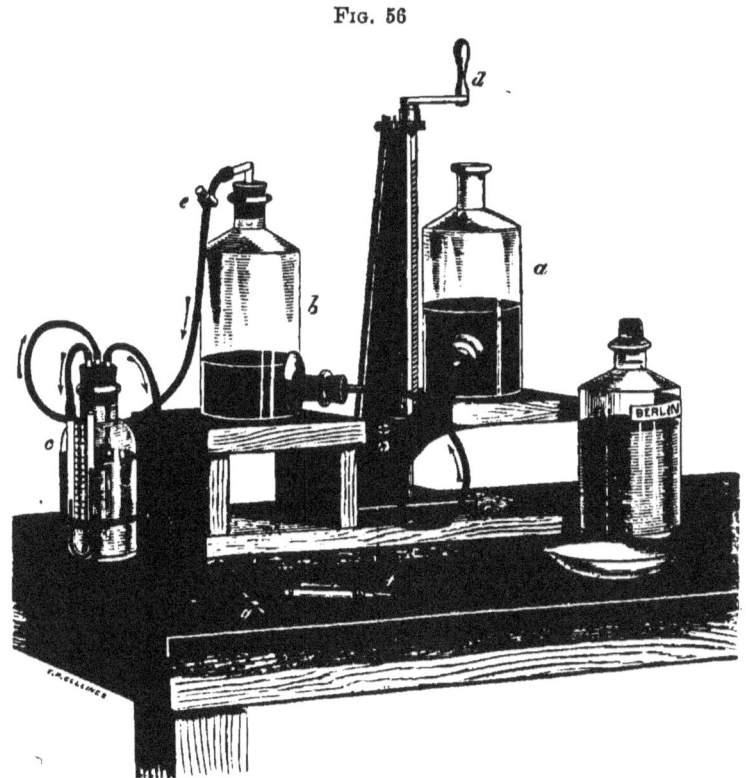

Fig. 56

Ludwig's mercurial-pressure apparatus for injecting lymphatics

a, reservoir-bottle containing mercury ; *b*, pressure-bottle into which this tends to flow ; *c*, injection-bottle containing solution of Berlin blue, connected with the pressure-bottle by one indiarubber tube, with the cannula *f* by a second, and with a small manometer by a third ; *d*, handle of screw, by turning which the stage on which the bottle *a* rests is raised or depressed, and the pressure increased or diminished in *b* ; *e*, screw clip (opened) ; *g*, spring clip (closed)

One of the best tendinous structures to choose for the purpose is the fibrous aponeurosis covering the tendon of the quadriceps extensor femoris of the dog. Two sets of lymphatics are here met with—one in the substance of the tendon,

o

consisting for the most part of vessels arranged conformably with the direction of the fibres and connected at intervals by transverse branches, so as to form elongated and oblong meshes; and a superficial one in the areolar sheath which covers the aponeurosis, consisting of vessels forming a close plexus with polygonal meshes. The latter plexus should first be attempted. Both tube and cannula being completely filled with the injection-fluid to the exclusion of air-bubbles, the clip g is closed, and the cannula is inserted obliquely for half an inch or more into the areolar sheath, care being taken not to discolour the surface of the tissue with the injection-mass. The cannula is then slightly withdrawn and the clip is removed. By turning the handle d, and thus raising the bottle a, the pressure is put on to about an inch of mercury, as indicated by the gauge attached to the

FIG. 57

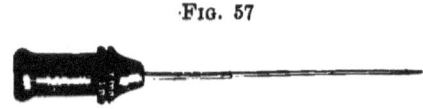

Very fine perforated steel needle for injecting the lymphatics of a part

injection-bottle. If the insertion of the cannula have been fortunate, the fluid will almost immediately begin to pass into the lymphatic plexus, but should there be no result the pressure may be gradually raised to about two inches; higher than this it is not as a rule advantageous to go. If there is still no result the cannula may be pushed a little further in the sheath, and perhaps moved a little to one side or the other in the hope of thus rupturing a lymphatic and gaining an entrance into the plexus. Should these and other devices which experience may suggest still fail, the clip must be replaced and another insertion tried elsewhere. It very frequently happens that the injection which escapes from the end of the cannula, instead of passing into the lymphatics, forms merely a bulla of extravasated fluid in the interstices of the tissue. This can sometimes, by passing the handle of a scalpel over it with moderately firm pressure, be induced to find its way into the absorbent vessels, but if not the cannula must be withdrawn and re-inserted as before.

For the lymphatics in the fibrous substance of the aponeurosis the cannula must be inserted obliquely into the tendinous tissue, and the injection forced in with the same precautions. The pressure may, if necessary, be raised somewhat higher, for, owing to the firmness of the tissue, there is less liability to extravasation.

For displaying these injected preparations they may, if injected with Berlin blue, be first placed in spirit to remove all water and precipitate the colouring matter in the vessels, the process being completed by putting the injected part into absolute alcohol, after which sections may be cut, placed in turpentine and mounted in xylol balsam. Another method, and one which succeeds very well, especially with the alkanet injection, is to stretch the injected aponeurosis over a ring of cork and allow it slowly to dry by exposure to the air. When completely dry the injected part may be at once mounted in glycerine. By this mode of proceeding the injection is, as pointed out by Bowditch, rendered more complete, for the fluid which may have been extravasated in the interstices of the tissue is apt to become drawn into the lymphatic vessels to supply the place of the watery fluid which becomes lost by evaporation.

Preparations injected with alkanet or asphalte must not be put in turpentine, but always mounted in glycerine.

Although in many cases it is better to use an apparatus of the kind above described, which enables the pressure which is being employed to be exactly estimated, nevertheless, with a little experience, the lymphatic vessels, especially those in the firmer tissues and organs, may often be injected with great success by using simply an ordinary subcutaneous syringe provided with a fine cannula, driving the injecting fluid into the tissue by pressure upon the piston. It is true that extravasations are very apt to be produced opposite the point of the cannula, but these can often be utilised in the manner before mentioned by gently pressing on them and endeavouring to induce the injection to pass into the lymphatic vessels.

In rare cases a vein is pierced by the cannula, and the system of blood capillaries of the part is then apt to become filled, but both the vessels themselves and the meshes they form are much smaller than the lymphatic capillaries, and a knowledge of their general appearance and mode of arrangement in the particular tissue will prevent any error from arising in this way.

Injection intra vitam.—The lymphatics of the diaphragm may be injected with Berlin blue during life :—A young rabbit is chosen, and enough chloral hydrate is injected under its skin to anæsthetise it completely (about 5 cubic cent. of a 20 per cent. solution will suffice). The skin of the belly is then cut through for a couple of inches close to the ensiform cartilage, the edges are held aside by an assistant, and the muscular wall having been pinched up, a cannula is passed obliquely into the cavity of the peritoneum at its upper part, due care being taken to avoid the liver and stomach. About five cubic centimetres of a saturated solution of Berlin blue, previously warmed, is now injected through the cannula, which is then withdrawn and the animal put aside in a warm place. After four hours, during the whole of which time it remains under the influence of the chloral, it is killed by bleeding. The abdomen is then opened, and the viscera having been drawn aside, the under surface of the diaphragm is exposed, and the blue which covers it is washed off by a stream of water. If the experiment has been successful, it will be found that the whole network of lymphatics of the central tendon is filled with the blue fluid ; for this, assisted by the constant respiratory movements of the diaphragm, has passed from the peritoneal cavity directly through the open stomata into the lymphatic vessels. The tendon may be cut out and placed in alcohol, and eventually, after passing through turpentine, mounted whole in xylol balsam, between two glass plates, and used for examination with a low power of the microscope.

THE SYNOVIAL MEMBRANES

These structures, which are to be regarded as free surfaces of the ordinary areolar tissue and present no essential differences in structural appearance from this, may be prepared for microscopical examination by the same methods. The preparations which are of greatest value are those stained with nitrate of silver.

Preparation by silver method.—Since, as is always the case with the silver method, the parts should be as fresh as possible, and since, moreover, it is convenient to have large joints to work with, a neat's or sheep's foot should be procured from the butcher's, unless a freshly amputated human limb is available. In the foot all three kinds of synovial membranes may be found and prepared. The mode of silvering the synovial bursæ is quite simple, and need not here be detailed; the preparation of the vaginal synovial membranes was described under Connective Tissue (p. 106), this being taken as typical of the structure of that tissue; and the preparation of the synovial surfaces of the joints was given under Articular Cartilage (p. 117). But in the last-mentioned place nothing was said as to the mode of demonstrating the synovial membrane proper, for we had there to do only with the cells and cell-spaces of the cartilage, and the transitions met with between these and the cells and cell-spaces of the synovial membrane. The appearance presented by the membrane itself is shown in surface sections made from the inner surface of the capsule of the joint. It will be seen that in the ox the cells, or rather, in the silvered preparations, the white cell-spaces, form a close irregular network by the union of their processes; in fact, so completely have the cells become extended into branches that without staining the nuclei it is difficult in many cases to make out the situation of the cell-body.

In the human synovial membranes this is not the case; in fact, the appearances are quite characteristic of ordinary areolar tissue. The arrangement of the corpuscles into epithelioid patches is not infrequent, but there is no continuous epithelioid covering, as in the serous membranes. Moreover, the lymphatics, which are so numerous in serous membranes, are not to be seen in the synovial membranes, although blood capillaries are present and in many places approach close to the surface.

Blood-vessels.—To study the characteristic arrangement of the blood-vessels of the synovial membranes and to show the *circulus articuli vasculosus*, a preparation must be made from one of the joints of a limb that has been fully injected. Surface sections are then made of the transitional region where the synovial membrane terminates on the cartilage, and including also a part of the membrane. They are mounted in the usual way in xylol balsam, without being stained.

Haversian fringes.—Finally the Haversian fringes, with their secondary processes, may be prepared. They may be examined fresh in salt solution, and may also be obtained from the joint which was stained with nitrate of silver. To find them it is best to immerse the joint in fluid, for by this means they are floated up, and may then be snipped off and mounted.

CHAPTER X

THE SKIN, HAIRS, AND NAILS

Sections of the skin.—Portions should be selected for examination from different parts of the body; the palms of the fingers or toes, the scalp, and a piece from some part of the general surface, *e.g.* the extensor surface of the forearm and the back of the trunk. The skin of the scrotum may also be prepared, to show the bundles of plain muscular tissue in the subcutaneous tissue or dartos, and a small piece of the ala of the nose, for the sake of the well-marked sebaceous glands, which open into the follicles of the minute hairs found in this situation.

The following method of hardening the tissue may be employed :—A small piece only is removed, being obtained with as little of the subcutaneous tissue as possible adhering to it. At the same time, if it is desired to examine the larger sweat-glands, this tissue must not be removed too freely, since those bodies extend down into it. It will be found that the fresh skin has a tendency to curl in at the edges; this should be prevented by pinning the piece out with glass pins or hedgehog spines on a piece of cork. The latter is then inverted into a vessel containing saturated solution of picric acid. Corrosive sublimate and 75 per cent. spirit also answer well. After having been two or three days in the picric acid the piece of skin is transferred to spirit containing lithium carbonate, which is changed until the excess of picric acid is extracted. The sections may be cut either by the freezing or the paraffin method, preferably the former. They should be vertical to the surface, but a piece of the scalp must be embedded so as also to be cut tangentially to the surface. The piece from the palm of the finger is to be cut in two directions, viz. (1) across and (2) parallel with the ridges formed by the papillæ. It will be easier to cut the sections if the razor is made to travel from the corium towards the epidermis rather than in the opposite direction. The pieces of skin which contain hairs should be so embedded as to carry the plane of the vertical section in the direction of inclination of the hairs, so as to gain a view of the hair follicles along their whole extent, and to show the *arrectores pili* if possible.

The sections are in each case to be placed in carmalum or logwood solution, and when sufficiently stained, to be washed in water and transferred successively through alcohol and oil of cloves to balsam in the usual way. Heidenhain's method (p. 20) also gives good results.

If hardened in alcohol or sublimate the preparations may be much improved, so far as the exhibition of the epidermis is concerned, by being placed in a saturated solution of picric acid for about half an hour and then thoroughly washed with distilled water before being stained, or in alcohol containing picric acid after being stained. In this way the horny parts acquire a bright yellow colour, which contrasts strongly with the violet staining of the Malpighian layer. Owing to the number of nuclei which become stained in it, this part of the epidermis is much darker than the papillary part of the corium which is in contact with it. In the papillæ of the skin of the finger the tactile corpuscles may be sought for. They are generally situated quite near the apex of a papilla, and to see them well it is important to cut the papillæ exactly vertically, so as to include their whole length. In sections which have been made somewhat obliquely, and in sections cut parallel instead of vertical to the surface, the transversely or obliquely cut papillæ appear as round or oval islands in the midst of the deeper cells of the epidermis. The denticulated appearance presented by most of the cells of the Malpighian layer, and the intercellular channels can readily be seen with a high power immersion objective.

Of the sections made from the skin of the finger one or two of the thinnest, after being stained with picrocarmine, may be mounted in glycerine. The fibrous-looking tactile corpuscles can generally be made out better in these than in the preparations which have been mounted in balsam.

Blood-vessels.—To show the arrangement of the bloodvessels, sections of skin from a limb which has been minutely injected may be made and mounted in balsam by the usual process. These must either be left entirely unstained or the staining must be slight. The sections will generally include clusters of fat-cells, with their vessels.

Hairs.—To examine a hair, all that is necessary is to place it on a slide in a drop of water, cover with a thin glass, and examine with a moderately high magnifying power. By careful focussing the cuticular scales can be made out on the surface and at the edges of the hair, especially on the small hairs of the general surface of the body. The medulla is often absent in hairs of the head, but may generally be found in those of the beard and whiskers. Many of the black particles which are seen in a hair by reflected light, especially in the medulla, are merely small globules of air in the interstices of the tissue. That this is so may be proved by cutting off the light which comes from the mirror of the microscope and viewing the object by reflected light, only a moderate power being used. The black particles, if really due to the presence of air, will then appear silvery white, just as in the parallel case of the air which fills the lacunæ in a section of hard bone.

It will be useful to compare the appearances presented by human hair with those exhibited by the hairs of some of the common domestic animals. These are many of them characterised by the regular arrangement of the medulla (this is nearly always present in the hairs of quadrupeds), which forms different patterns in different kinds of animals, so that the species to which the hair belongs may often be determined.

The fibrous part of a hair can be broken up into its constituent fibres and cells if it be first steeped for a time in strong sulphuric acid.

The relative proportion of the three constituent parts of a hair to one another is best shown in transverse sections.

To obtain sections of hairs, the simplest plan is to tie a number together and dip the bunch into strong clear gum, and when this has thoroughly soaked in amongst the hairs to remove the bunch, and either let the gum dry and harden by exposure to the air or plunge it into spirit containing one-sixth water, by which in a few hours the whole mass is rendered hard. Sections are then made with a microtome provided with a sharp plane-iron, and are mounted in glycerine.

Hair-follicles and roots of hairs are seen in the sections of skin, especially those of the scalp.

Nails.—The nails are studied by means of vertical sections made both longitudinally and transversely. The finger (or toe) should if possible be previously injected, and the nail with the matrix and surrounding skin having been removed, may be hardened either in spirit or sublimate or by picric acid and cut frozen after being soaked with gum. When ready for cutting, the piece should be bisected longitudinally ; and from one of the halves longitudinal sections (which need not be very thin) are taken along the whole length, to show the general relation of the nail to its matrix and to the epidermis. These may be stained or left unstained; they may be mounted either in glycerine or, if injected, in balsam ; they are intended chiefly for examination with a low power. The other piece is to be placed on the microtome in such a way that the laminæ, which in the matrix represent the papillæ of the skin, are cut transversely. The sections must be as thin as possible, and stained either with picrocarmine or with logwood, or with carmalum followed by alcoholic solution of picric acid.

THE SKIN 203

But, owing to the substance of the nail being so much harder than the subjacent matrix, it is very difficult to get both parts equally thin. They can, however, be got of much the same degree of hardness by means of the gum-alcohol method. The piece to be cut, which should be quite small, is placed in syrupy solution of gum and left for two or three days ; it is then transferred to a mixture of spirit with one-sixth of its volume of water. After a few hours the gum, which has penetrated into the substance of the tissue, will be hardened throughout, and the mass can be fixed in a microtome and cut in the desired direction, the plane which is used being wetted with some of the same spirit-mixture. Strong spirit should not be used, since the gum is entirely dehydrated by this, and becomes so hard as to turn the edge of the knife. The sections are transferred from the spirit to water, which dissolves out the gum ; when quite free from this they are stained and mounted as before.

Nerves to the skin and its appendages.—The nerve terminations in the epidermis, in the hair-follicles, in the sweat-glands, and in other parts of the skin, may be studied either by Löwit's gold chloride method (p. 159) or by Golgi's silver chromate method (p. 152), or by Ehrlich's methylene blue method (p. 160).

Lymphatics.—The lymphatics of the skin may be studied either in preparations made by the method of interstitial injection (p. 192) or in silvered preparations. The latter can be best obtained from the skin of a small animal, such as the mouse. A piece of the skin is stretched over a Hoggan's ring (fig. 55), and, the hair having been shaved off, it is immersed in 1 per cent. nitrate of silver solution for 15 minutes ; then washed with distilled water and exposed to the light. When brown it is placed in spirit to dehydrate it, then in clove oil, and finally a piece is cut out and mounted whole in xylol balsam.

Development of hairs and skin-glands.—To study the development of hairs and of the skin generally, sections of parts of fœtal animals, which have been hardened in 0·2 per cent. chromic acid or picric acid (saturated), may be employed. Many stages in the development of hairs are seen in such sections from the new-born rat. The tissue should be stained in bulk and embedded in the usual manner in paraffin. The sections must be fixed to the slide (p. 37), and if found to be over-stained the excess of stain can then be removed by the cautious use of acid alcohol.

CHAPTER XI

THE HEART

The cardiac pericardium.—The pericardium which covers the surface of the heart is prepared by the same methods as other serous membranes. Of these the only one which need here be described is that by nitrate of silver. This is as follows :—In an animal which has just been killed the thorax is freely opened, and the pericardium having been cut open, the base of the heart is secured by a tape ligature, the great vessels being then cut beyond the ligature, and the organ removed without allowing its surface to be smeared with blood. A part of the surface is now brushed firmly with a soft camel-hair brush moistened with 2 per cent. nitrate of soda solution, with the object of removing the superficial layer of epithelial cells. The heart is next rinsed with the same, and nitrate of silver solution is poured over the whole surface, and allowed to be on it for three minutes, after which the organ is rinsed in distilled water, and finally placed in spirit, in the sunlight. When sufficiently browned it is removed from the window, and left for some hours until the surface is hardened by the alcohol. Surface sections are then made of both unbrushed and brushed parts, and after being passed through clove oil are mounted separately in balsam. The unbrushed specimens will show the epithelial layer ; the brushed ones should exhibit the subjacent connective tissue, with its cell-spaces, lymphatics, blood-vessels, and nerves.

The muscular substance of the heart is studied in teased preparations and in sections. For the teased preparations the heart of a young animal should be chosen, since in these the fibres separate more readily into their constituent cells. A small shred is placed in a solution of picrocarmine for ten days or more; it is then broken up in water as minutely as possible, and glycerine added and the preparation covered. Numerous little fragments of varying shapes will be found scattered over the preparation. On careful examination it will be apparent that each possesses a nucleus. These little fragments of the cardiac muscular tissue, which have the characteristic indistinct striation of that substance, are the cells which by their union end to end form the fibres.

To show the arrangement of the fibres, and the interstitial tissue and vessels, a piece of the muscular substance is to be placed in strong spirit. In two or three days it will be ready to embed and cut. Sections are to be made both parallel with and across the direction of the fibres; they are to be stained with hæmateïn and mounted in balsam.

The endocardium.—To display the endocardium the silver method again comes into requisition. The part of the lining membrane which covers the septal wall of the right ventricle is the best to prepare, on account of its relative smoothness. The right ventricle is opened in a fresh heart, and the outer wall removed entirely, and then a large piece of the smoothest part of the exposed surface of the septum is sliced off with a razor. A part only of the endocardium of the detached piece is brushed, as in the case of the pericardium, and the whole is then washed and treated with silver solution. After three minutes it is put into spirit as before, and when browned and hardened, surface sections are cut and mounted in balsam.

In addition to these silvered preparations, the endocardium should be examined in the fresh state. This is done by dissecting off a piece of the membrane in salt solution and examining it both with and without the addition of acetic acid. Other portions may be teased out with a view to the demonstration of the elastic and muscular tissue. The methods for making and preserving these preparations are the same as were employed for showing the structure of the coats of the blood-vessels, to the description of which the student is referred (p. 164). It may be noted that in some animals—the sheep, for instance—the peculiar large cubical or oblong cells which form, in series, the fibres of Purkinje will be found in the endocardium. They are about the size of fat-vesicles, which are also found in the endocardium in this animal; but the two can hardly be confounded, for the cells forming Purkinje's fibres have a clear or slightly granular binucleated central portion which does *not* strongly refract the light, and a striated circumference, which is apparently continued into that of the neighbouring cells; whereas the fat-cells, although they may also occur in rows, and may be of much the same size as the cells in question, present, by virtue of their strong refracting power on light, a totally different appearance.

Lymphatic system of the heart.—If the fine cannula of a hypodermic syringe filled with Berlin blue solution is stuck into the muscular substance of the fresh heart at any part, and the fluid is forced out at the point, the injection will pass freely into the lymphatic interstices between the muscular fibres, and if the tube is inserted near the outer or inner surface, will find its way into the lymphatics of the pericardium or endocardium, which can in this way be readily displayed.

Blood-vessels.—The blood-vessels of the heart are to be studied in sections from an injected preparation. The nerves may be stained by the silver chromate, or by the methylene blue method, but they are somewhat difficult to demonstrate satisfactorily.

CHAPTER XII

THE LUNGS

The pulmonary pleura.—The serous membrane which covers the surface of the lungs as well as that which lines the wall of the thorax is prepared by the silver process. For the pulmonary pleura, the lungs of a small animal that has just been killed are to be removed entire, and moderately distended with air through the wind-pipe, the bronchi being then tied and the two lungs separated. One is rinsed for a moment in distilled water, and a little nitrate of silver solution is allowed to flow over the surface; after the lapse of a minute this is washed off again with distilled water, and the organ is then immersed in a beaker of spirit and exposed to the light.

The surface of the other lung is to be firmly brushed with a wet camel-hair pencil, to remove the epithelial cells of the surface before treating it with the silver solution. This may be suffered to remain longer in contact with it than with the other lung (5 minutes); in other respects the treatment is similar. Both preparations are left in the light until they appear sufficiently stained, after which they are to be placed on one side in the spirit for twenty-four hours. They will then be sufficiently hard to render it possible to shave off a thin slice from the surface. The sections so made are to be mounted in balsam, with the outer surface uppermost.

The costal pleura is to be prepared *in situ* after the removal of the lungs and heart. That of one side may be brushed, the other not; on the latter the silver solution is, as before, to be allowed to remain a shorter time than on the brushed part, where the fluid has to penetrate into the lymphatic vessels, and into the substance of the tissue. The whole thorax, if small (*e.g.* mouse), may then be cut off from the rest of the trunk and exposed in spirit to the light ; or, if too large to do this conveniently, a piece only of the thoracic parietes on each side is to be removed and pinned out on to a cork, which is then placed in a vessel of spirit in the sunlight. When stained, pieces of the membrane must be carefully dissected off, without pulling upon or injuring the tissue in any way, floated upon a slide, the excess of spirit poured off, all creases removed from the membrane, which is then allowed to dry, and finally mounted in balsam.

The lung tissue.—The structure of the lungs themselves is best shown by means of sections. The tissue is hardened in the following way :—

The organs having been removed from the chest of a recently killed animal, care being taken not to scratch their surface with the broken ends of the ribs, a glass cannula is tied into the end of the trachea (or into either bronchus). The cannula is then connected by an indiarubber tube with an injection-bottle, which is filled with a weak solution of chromic acid (0.2 per cent.) ; this is made to flow into the lungs so as to distend them moderately. The trachea or bronchus is now tied up, the cannula removed, and the lungs are immersed in a large quantity of a solution of chromic acid of similar strength. After a few hours the fluid is changed, and the organs are cut into pieces, to enable the fresh fluid more readily to penetrate. After a day or two more in this the pieces are placed first for twenty-four hours in weak spirit, and then in strong spirit, which may again be changed once or twice. Picric acid or corrosive sublimate (saturated), formol (5 per cent.), or bichromate of potash (3 per cent.), may be used in place of chromic acid.

The pieces should be stained in bulk with dilute hæmateïn or with carmalum, or by Heidenhain's method, and

P

embedded in paraffin. Sections may be made both across and along the course of the main bronchial tubes, and, after going through the usual processes, mounted in balsam.

These stained sections of lung may be first examined with a moderate power, but afterwards a power of 400 or 500 diameters should be employed, in order to see the details of structure ; the ciliated epithelium, muscular layer, and cartilaginous plates of the bronchial tubes, with the mucous glands, nerves, lymphatics (seen in section as mere clefts), and patches of lymphoid tissue in their walls ; the branches of the pulmonary artery accompanying them ; the mode in which the terminal air-tubes dilate into the infundibula ; the air-cells or alveoli, almost covered with a network of capillaries, which are seen also on the septa between the alveoli, projecting first into one and then into the other of two neighbouring air-cells. Where they run vertically the capillaries appear in optical section as circular spots, looking not unlike nucleated cells. But the excessively delicate epithelium of the air-cells cannot be well seen in these preparations, for the epithelium cells remain almost unstained, and it is not easy to differentiate their nuclei from those of the closely subjacent capillaries.

Epithelium lining the air-cells.—In order to demonstrate the epithelium cells nitrate of silver is made use of, but the mode of proceeding is somewhat different from that ordinarily employed. A gelatine mixture is made by taking ten grammes of gelatine, and, after soaking it in cold distilled water, melting it, and adding if necessary more warm distilled water until the mixture measures 100 c.c. A decigramme of nitrate of silver is dissolved in a little distilled water and added to the gelatine, and the mixture is transferred to an injection-bottle, which is kept warm over a water-bath. The injection-bottle is provided with an indiarubber cork and two glass tubes, one of which serves to blow air into the bottle and thus raise the pressure, and the other to conduct the gelatine by means of an indiarubber tube to a cannula tied into the trachea or into a bronchus. An animal—preferably a young one—having been killed by bleeding, the lungs are removed, the glass cannula is tied into the trachea, and enough of the gelatine mixture is injected into the lungs to distend them pretty completely. The trachea is now tied and the cannula removed from it. The lungs are then put aside into a cold place until the gelatine within them has fully set, when sections, which should be as thin as possible, are made with a razor, either not wetted at all or with distilled water only. The sections so obtained are placed on a slide in glycerine, covered, and exposed to the light. As soon as they seem sufficiently stained they may be examined with as high a power as possible, for the purpose of making out the silver-lines between the epithelium cells.

Blood-vessels of the lungs.—The pulmonary vessels are to be injected and sections made of the injected lung. The red gelatine injection may be used; this and everything else is to be got ready in the same way as for the injection of the aortic system, but the trachea must be clamped in the neck before the chest is opened, in order to prevent the lungs from collapsing. The arterial cannula is of course to be passed through the *right* ventricle and tied into the pulmonary artery instead of into the aorta. The gelatine injection is now set flowing through the pulmonary vessels, the lungs being still distended with air; the pressure is then raised in the injecting-apparatus to about three inches of mercury. The left ventricle is first slit, to let the blood out of the pulmonary system, and then clamped, to prevent the escape of the injecting-fluid, which is allowed to pass until it is thought that the vessels must all be completely filled. The trachea is now opened so that the lungs collapse, but they are brought back to their previous condition by the injection of warm 3 p.c. potassium bichromate solution into them through the windpipe. The trachea and the base of the heart are then ligatured, and the whole is left to cool. The lungs are then cut out and placed in 3 p.c. potassium bichromate; after a few days sections (not too thin) may be made by the freezing method and mounted in balsam. If it is desired to stain the tissue somewhat, so as to show the general structure of the lung as well as the arrangement of the blood-vessels in the same preparation, this can be done by hæmateïn.

The injected lung may also be distended, while the injection is still flowing, with melted paraffin, and when cold placed in alcohol, the sections being cut by the paraffin method.

Larynx and trachea.—The trachea and larynx are hardened in 0·2 per cent. chromic acid, the hardening being completed by spirit, as with the lung. If not previously stained in bulk the sections, which may be longitudinal of the cartilaginous part and transverse of the posterior membranous part, are to be stained with logwood or carmalum and mounted in balsam.

Blood-vessels, lymphatics, and nerves of trachea.—In addition to these sections flat preparations showing the blood-vessels, and others showing the lymphatics of the mucous membrane, may be made. The former are got from any animal that has been injected entire, the mucous membrane being dissected off and mounted in balsam. The lymphatics are readily filled by the puncture method by sticking the point of the injecting cannula into the mucous membrane, and forcing in a little Berlin blue or alkanet-turpentine. It will hardly ever fail in finding its way into the numerous lymphatics of the mucous membrane. The injected portion is dissected off and mounted in balsam or in glycerine. The medullated nerves may be shown by merely exposing the fresh mucous membrane to vapour of osmic acid (Stirling); their fine terminations by the silver chromate or methylene blue method.

Teased preparations to show the separated epithelial cells have already been described (p. 94).

CHAPTER XIII

THE MOUTH AND PHARYNX

Mucous membrane of the mouth.—Portions of the lining membrane of the mouth are best prepared by being pinned out upon a cork and placed in strong spirit or absolute alcohol. Sections of the cheek or lip may also thus be readily prepared.

THE TEETH

Sections of hard tooth.—No preparations exhibit the general structure of the teeth better than these. A slice cut with a fine saw from the hard tooth is ground down first on one side and then on the other, until a thin section only remains, and this is mounted in hard Canada balsam in such a way that the air still remains in the dentinal tubules, the lacunæ of the cement, the interglobular spaces, and other minute cavities that may be present. The preparation is similar to that of bone, but presents greater difficulty. Such specimens may advantageously be purchased, for their preparation involves the expenditure of a large amount of time and labour ; unless the use of a lapidary's wheel can be obtained, when the process is much facilitated. They should in every case be studied first with a low power, and afterwards with a high power objective.

Von Koch's method.—Another method of preparing sections of hard teeth with the soft parts preserved *in situ* is that devised by v. Koch for obtaining thin sections of calcareous sponges and corals, and applied to the histology of the teeth by Weil. The tooth is preserved by being placed in absolute alcohol for a few days; other fixing fluids may be used: *e.g.* Hermann's or Flemming's, but they should be followed by alcohol. It is stained in bulk by being placed for several days in picrocarmine, or in carmalum. It is then dehydrated by alcohol of gradually increasing strength, finishing up with absolute alcohol, and being subsequently placed in chloroform. From this it is transferred to a strong solution of hard Canada balsam in chloroform, in which it also remains for several days. When removed it is allowed to dry slowly, until the Canada balsam has become hard throughout. Sections are now made with a fine saw, and are ground down with the aid of a lapidary's wheel, until they appear thin enough; they are then mounted in Canada balsam.

Sections of softened tooth.—But, in addition to the facts which the hard specimens will show, various others may be made out in teeth which have been softened by immersion in an acid. The methods of decalcification are the same as for bone (pp. 125 to 127).

The acid generally used when the structure of the (decalcified) dentinal substance only is to be investigated is nitric or hydrochloric. A nearly saturated solution of common salt containing 10 per cent. of either of these acids may be employed, and the tooth is steeped in this until entirely soft, after which it is placed in the salt solution without acid. After several changes into fresh quantities of this, it may be preserved in spirit. Sections are to be made in planes both parallel with and across the direction of the dentinal tubules.

Dentinal sheaths.—To show the sheaths which line the dentinal tubules a piece of such softened tooth is transferred to strong hydrochloric acid (contained in a watch-glass, which is covered by another, inverted) and left for about an hour, after which time all that will be found is a tenacious soft mass occupying its place. If some of this mass be removed with a small pointed piece of wood, placed on a slide, covered, and examined, it will be found to consist wholly of fine tubular threads—the dentinal sheaths. These being composed of a substance which resists the solvent action of strong hydrochloric acid longer than the other animal tissues, remain for a time visible after the rest of the dentinal substance has disappeared.

Soft tissues of the teeth.—But to study the soft tissues— that is to say, the pulp and odontoblasts with the processes which these send into the dentinal tubules—we must, as in the parallel case of bone, employ a reagent which, whilst softening the hard parts, at the same time preserves and hardens the soft parts. Phloroglucin-nitric acid (p. 127) or picric acid may be employed. If the latter, the freshly extracted tooth is placed in a saturated solution of the acid containing two parts per cent. of hydrochloric acid, and crystals of picric acid are from time to time added as required, the solution being stirred frequently with a glass rod. It is well to break or cut open the tooth first, so as to expose the pulp cavity, if this can be done without disarranging the contents. When softened throughout—this can be tested by attempting to pass a fine needle through it—the tooth is placed in spirit (containing lithium carbonate), which should be changed until it ceases to become coloured by the excess of picric acid. All that remains to be done is to embed in paraffin by the usual method, fix the sections to a slide and stain with hæmateïn or picrocarmine or some other dye, mounting either in glycerine jelly or in xylol balsam.

Study of decalcified teeth 'in situ.'—Still more instructive preparations are obtained by softening a portion of the lower jaw with the teeth *in situ*, and making sections through the whole structure. It is best to take the jaw of a small animal—a rat, for instance. The flesh having been cleared away, the softening is effected with picric-hydrochloric acid or with phloroglucin-nitric acid ; and then, after due washing with lithium carbonate spirit, the piece is embedded, cut, stained, and mounted as before.

Besides showing the teeth and the way in which they are inserted into the lower jaw, the structure of this bone is itself well demonstrated. At the lower part the constantly growing incisor, which extends in the rat below the molars to the back part of the jaw, exhibits the large elongated odontoblasts of a developing tooth, with their well-marked dentinal processes, which in some parts project like harp-strings across a small space, produced by shrinkage, which intervenes between the cells and the dentinal substance. It will be remarked, also, that in these teeth the most newly formed layer of dentine becomes, especially near its junction with the older parts, very intensely stained by the logwood. This is the case with all teeth which are still in process of development. Carmine has not the same action.

Development of the teeth.—For the study of the development of the teeth sections are made of the jaws of embryos and young animals; perhaps the most convenient to choose are fœtal and new-born rats. Sections of their jaws exhibit not only the mode of development of the teeth, but also that of the hair, the bone of the lower jaw (which ossifies in the connective tissue around Meckel's cartilage), the tongue, and many other parts. The preparation is as follows:— The fœtuses or young animals are decapitated, and the heads dropped into 0.2 per cent. chromic acid or Flemming's fluid. After a week's time, during which the liquid is frequently stirred, they are transferred to weak spirit containing lithium carbonate, and in a day or two to strong spirit. After being in this for two or three days they are ready for cutting. Either the lower jaw is removed and embedded separately, or the whole head is embedded, and both jaws are cut simultaneously. The sections are to be stained with logwood, carmalum or picro-carmine, unless the preparation has been stained in bulk. The stay in chromic acid may not have been long enough to remove all the earth from the partly developed bones and teeth, but what still remains is so small in amount that it will not prevent a thin section being made. The earlier stages in the development of the teeth may be perhaps seen in the molar region; the later stages, comprising the development of the dental tissues, especially the

THE TONGUE 217

dentine and enamel, may be studied in the much more advanced incisors, which, as just pointed out, extend backwards in these animals through the greater part of the length of the jaw.

THE TONGUE

Sections.—Small portions of this organ from different parts are hardened in 3 per cent. bichromate of potash (fourteen days), and subsequently in spirit, and are embedded, so as to be cut vertically to the surface of the mucous membrane. The sections are stained with hæmateïn, and mounted in balsam. A double staining with picric acid and logwood or with picro-carmine may also be employed in the same way as with the sections of skin (p. 200). The stratified epithelium, the papillæ of the mucous membrane, and the arrangement of the muscular fibres, as well as the mode of termination of the superficial muscular fibres in the connective tissue of the mucous membrane, may be studied in these sections. Some of the sections include mucous glands, which differ in appearance according to their condition at the time of death. If they had not been recently stimulated by the ingestion of food or otherwise, the cells of the gland will still be filled with mucus, as shown by their swollen appearance, and by their becoming strongly stained by the logwood; whereas, on the contrary, if they have recently discharged their secretion, the cells will be granular and almost colourless, and there will be indications of mucus in the lumina of the ducts.

Besides the mucous glands others may be seen in the neighbourhood of the papillæ vallatæ, which secrete no mucus, and consequently do not present the above differences of staining. Many of their ducts, if the section pass in the direction they take, will be found to open into the fossæ of the circumvallate papillæ, or, at least, near those parts in which taste-buds have been found. The taste-buds themselves may possibly be seen in the epithelium on the sides of the papillæ vallatæ, and also in that which covers the mucous membrane near the root of the tongue on each side. The description of other modes of preparing them will, however, be deferred until the organs of special sense are treated of.

Vessels of the tongue.—Sections should also be made of an injected tongue. These will show not only the numerous vascular loops in the more obvious papillæ of the mucous membrane, corresponding in number with the microscopic secondary papillæ, but also the arrangement of the vessels in the muscular substance of the organ. The injected specimens are much improved by slightly staining them with logwood or eosin.

Palate and tonsils.—The soft palate and the tonsils may be hardened in the same way as the tongue, or, by immersion in strong spirit. The sections are stained and mounted in the usual way.

THE SALIVARY GLANDS

Sections.—These organs are prepared by placing small pieces of them as soon after death as possible in formol, sublimate, or picric acid solution, or in Flemming's or Hermann's fluids. After two or three days they are transferred to spirit, which is frequently changed. They may also be prepared by merely being placed in strong spirit or absolute alcohol for two or three days. The embedding, staining, and mounting may be effected in the ordinary manner, but bulk-staining by Heidenhain's method is especially to be recommended for these structures.

Physiological states of the cells.—The difference in the appearance of the salivary cells previous and subsequent to the state of secretory activity is best studied in the glands of two animals, one of which has been killed after some hours' fasting, the other a short time after administration of a small dose of pilocarpin by hypodermic injection. A small portion of each gland is prepared with picric acid for sections, and another is treated with vapour of osmic acid and afterwards macerated in water for a few days for teased preparations. Besides these, other small fragments are to be teased fresh in 2 per cent. salt solution.

Gland ducts and nerve terminations.—The ducts of the glands and their intracellular commencements in the alveoli are studied in sections of small pieces of fresh glands prepared by the method of Golgi (see p. 152). In these sections nerve terminations may also be observed.

CHAPTER XIV

THE ŒSOPHAGUS AND STOMACH

Sections of œsophagus.—The œsophagus is hardened for the preparation of sections by picric acid, 0·2 per cent. chromic acid, Flemming's fluid, formol, alcohol, or 3 p.c. bichromate of potash. After three or four days in any of the first three, the tissue may, as usual, be transferred to spirit. Before putting it into the mixture it should, if but a small piece be employed, be pinned out upon a piece of cork, so as to stretch it slightly and avoid folds. But if a tubular piece be available, this object may be effected more satisfactorily by distending the organ with the preservative solution through a glass cannula tied into one end, the other end having been secured by a ligature before the distension; the piece is then immersed in the mixture for a few hours, after which it may be cut open.

Heidenhain's bulk-stain gives good results with the œsophagus.

In cutting a piece of one of the membranous viscera it is well to place it, as a general rule, so that the sections shall be exactly transverse to the axis of the viscus, following, therefore, the direction of the circular muscular fibres and cutting the longitudinal across. When the direction of the section is known, it is easier to understand the appearances which the various parts present when cut. Sections cut parallel to the axis of the viscus, and taking therefore the direction of the longitudinal muscular fibres, may be made with nearly equal advantage, but oblique directions should be avoided.

Blood-vessels of the gullet.—The arrangement of the

vessels is best shown in a flat preparation. A small piece, obtained from a small injected animal, is transferred, without staining, from spirit to oil of cloves, and subsequently mounted in balsam, with the inner surface uppermost. Such a preparation is only useful for examination with a low power, but by this the arrangement of the vessels in the successive strata can be well made out. If too thick to be mounted as one specimen, the mucous membrane and muscular coat may be dissected apart and mounted separately. Vertical sections of injected œsophagus may also be made.

The stomach.—The stomach should always be prepared as soon as possible after death, for, in the first place, the columnar epithelial cells covering the inner surface soon become altered; and secondly, if digestion were proceeding in the organ at the time of death, the mucous membrane itself becomes attacked by the gastric juice in a very short time.

The abdomen, therefore, is to be opened as soon as the animal (a cat or dog) is dead, the œsophagus cut as near the diaphragm as possible, and the duodenum about two inches beyond the pylorus; the folds of peritoneum connecting the viscus to the liver and neighbouring parts are also severed, and the stomach is removed. If the organ is empty or if the contents are fluid enough to admit of being poured out through the pylorus, it may be prepared as a whole by distension with the hardening fluid, which may be corrosive sublimate, formol, picric or chromic acid, followed by spirit. The duodenal end is tied up and a glass cannula is fastened into the œsophageal end. This is connected by an indiarubber tube with a bottle containing salt solution. A second tube passes through the cork of this bottle, and by blowing through it the salt solution is forced into the stomach and washes out its contents. The salt solution is then replaced by the hardening fluid. When the organ is moderately distended with this the indiarubber tube is clipped, to prevent any of the liquid being forced back into the bottle by the contraction of the muscular walls of the stomach; the lower end of the gullet is then secured by a ligature, and the whole organ is immersed in a large receptacle filled with the same hardening fluid. After a few hours it should be opened and put into fresh fluid ; or if it is not desired to keep the whole of the organ, small pieces only from different parts are so transferred. If the stomach be too large to harden as a whole, or difficult to be cleared of its contents, small pieces may be cut out from the fresh organ, pinned out on a piece of cork or cake of wax, washed with salt solution, and then immersed in the hardening fluid. In two or three days the tissue is ready to be transferred to spirit. The sections, which should comprise all the coats of the organ, are to be stained with carmalum, or with hæmateïn and eosin, or with toluidin blue and eosin, and mounted in balsam as usual.

Heidenhain's bulk-stain may also be employed with advantage.

One set of sections should be taken longitudinally through the pylorus in order to show the manner in which the pyloric glands of the stomach become continued into the duodenum as the glands of Brunner. Other sections should pass through the junction of the œsophagus with the stomach.

Gastric glands.—For making out distinctly the structure of the mucous membrane and the character of the cells which occupy the gastric glands, it is better to take small portions of the mucous membrane only. With this object, small pieces of the fresh mucous membrane are taken from two or three distinct parts—one from near the pylorus, one from the fundus, and one from close to the œsophageal opening—and placed at once in absolute alcohol or in corrosive sublimate or in picric acid, to be followed by alcohol. When hardened they are embedded separately, and vertical sections, as thin as possible, are made. Some of these are to be stained with carmalum and picric acid and mounted as before in balsam, but one or two of the thinnest may be selected and mounted in glycerine. Others may be placed for twenty-four hours or more in picrocarmine solution. Others may be stained with toluidin blue and eosin, and yet other sections with Nicholson's blue No. 1, and mounted in glycerine. Lastly, a small fragment from each part may be put fresh into 1 per cent. osmic acid, and after 24 hours washed in water, embedded in gum, and cut frozen.

These different modes of staining bring out distinctly the differences between the various kinds of cells found in the glands. In the logwood preparations the parietal cells will be found stained rather less than the rest, whereas in the sections stained by carmine, eosin, and toluidin blue they are coloured more deeply. This is especially the case in the Nicholson-blue preparations, where the peptic cells are stained of a deep colour, whilst the other cells may remain almost uncoloured. Many animals have a special kind of glands near the cardiac orifice, differing from the fundus-glands in containing no parietal cells.

Horizontal sections.—Besides the vertical sections of the mucous membrane, others are to be made parallel with the inner surface, and therefore so as to cut the glands across. The sections will, of course, comprehend in succession first the mouths of the glands, then their necks, and finally the deeper parts. They are to be stained and mounted in the same way as the vertical sections.

Cells of the glands, isolated.—In addition to studying them in sections in this way, the cells of the glands may also be studied in teased-out preparations of the fresh mucous membrane made in serum or 0·9 per cent. salt solution. The cells will be more readily obtained separate if a small piece of the membrane is placed in one-third alcohol for twenty-four or forty-eight hours, but are apt to be somewhat altered, and the columnar cells of the general surface and mouths of the glands to become transformed into goblet-cells, by the swelling and escape of their contained mucus.

Blood-vessels of the stomach.—Both vertical and horizontal sections of an injected stomach may be made. This may be obtained from the animal which was injected entire : if it were a rat, the preparations are to be made from the pyloric half of the organ, since in this animal the cardiac end has a non-glandular mucous membrane with stratified epithelium like that of the gullet. The vertical sections need not be very thin; they are improved by being stained with alcoholic eosin, so as to become slightly coloured, before being mounted in balsam by the ordinary process. Instead of cutting horizontal sections a small piece of the injected stomach may, if from a small animal, be simply mounted flat with the inner surface uppermost, without staining.

Lymphatics.—An attempt may be made to inject the lymphatics of the gastric mucous membrane with Berlin blue, but the process requires considerable care and experience, since it presents unusual difficulties. If a successful result is obtained the injected portions are hardened in alcohol, and vertical sections, which may be tolerably thick, made and mounted in balsam.

Nerves.—The nerve plexuses and ganglia in the coats of the stomach may be shown by the methods recommended for the intestinal nerves (p. 230). The silver chromate method also serves to show the ductules which lead from the lumen of the fundus glands to the parietal cells.

CHAPTER XV

THE SMALL AND LARGE INTESTINE

Sections of small intestine.—Pieces of the small intestine are prepared in exactly the same way as the stomach, the gut being distended by the hardening fluid. After a few hours the intestine is opened and the fluid changed, and in three or four days the tissue is transferred to spirit, to complete the hardening. Three pieces of the small intestine are to be preserved in this way—viz., one from the commencement of the duodenum (this will probably have been included in the stomach preparation); a second from the jejunum; and the third from the ileum, including one of the patches of Peyer. The pieces may be obtained from a cat, dog, or rabbit, the contents of the intestine being first cleared out by gently squeezing the gut, or by forcing a rapid stream first of salt solution and then of the hardening fluid through each piece before tying up the further end. Instead of distending it with the preservative fluid, the gut may be opened and kept in an extended state by pinning it on a cork or cake of wax, which is then inverted into the fluid.

It is necessary, in order to see the structure of the villi, that the sections should be very thin so as to include not the whole thickness of a villus, but only a longitudinal slice, otherwise the epithelium on its surfaces interferes with the view of the internal structure. One piece must be so cut as to obtain sections across the villi, which is easily done with paraffin-embedding, the sections being fixed to the slide in the usual way.

Fat absorption.—For the purpose of studying the course

which fatty particles take in passing from the cavity of the intestine into the central lacteals of the villi, a rat is killed three or four hours after a meal containing fat. On opening the abdomen the lacteals in the mesentery should be found filled with chyle, and the cavity of the small intestine occupied by emulsified fat which is undergoing absorption. The intestine is opened at once, and two or three very small pieces of the mucous membrane are snipped off and treated with 1 per cent. osmic acid or osmic acid vapour for an hour. Another minute piece is quickly teased out with needles in a drop of serum or salt solution; a piece of hair is added, and the preparation is covered and examined. One of the portions treated with osmic acid is allowed to remain forty-eight hours in water, and is then broken up with needles and by tapping the cover-glass. Another is placed in gum, and when permeated with this sections are cut, frozen, and mounted in glycerine or glycerine jelly.

In the two teased preparations—serum and osmic—many of the columnar epithelium cells will be found to contain fatty globules of various sizes (stained black in the osmic preparation). Similar, but for the most part smaller, particles will also be found in some of the numerous lymph corpuscles which are set free from the retiform tissue of the mucous membrane by the process of teasing. In the sections the epithelium cells and the lymph corpuscles will be observed, *in situ*, in the same condition, viz. containing blackened fatty particles, and moreover the cleft-like central lacteal in the middle of each villus may be found to contain not only similar globules, but also lymph-cells in process of disintegration. Hence we infer that the fatty matters are first taken up from the cavity of the intestine by the columnar epithelium cells; that they are transmitted from these to the interstitial tissue of the villus, where they are taken up by amœboid lymph-cells, and that these convey them into the central lacteal. Where absorption is proceeding very rapidly, some of the fatty particles may be found in the interstices of the connective tissue of the villus, not included within leucocytes ; and it is probable that under these circumstances some of the fat may pass into the lacteals independently of those cells. This is especially the case in the dog, but in other animals, such as the rat and guinea-pig, the whole of the fat is usually included within either the epithelium cells or the leucocytes, and only becomes free within the lacteals. Similar observations may be made on a frog killed two days after being fed with lard.

Vessels of the small intestine.—The *blood-vessels* of the small intestine are studied in vertical sections of the injected gut. The sections may be stained with eosin.

The *lymphatics* (lacteals) may perhaps be seen in thin sections of the uninjected preparations as cleft-like spaces in the villi and in the substance of the mucous membrane, and surrounding the bases of the lymphoid nodules which make up the Peyerian patches. It is not an easy matter to inject those of the mucous membrane, although the larger plexuses of the submucous and muscular coat can be more easily demonstrated in this way.

Nerves of the intestinal wall.—The nerves of the intestinal canal form a very interesting subject of study, comprising some of the closest and most richly gangliated plexuses of fibres which are met with in the sympathetic system. They may, moreover, by the use of the chloride of gold, the chromate of silver, or the methylene blue methods, be shown in all parts of either the small or the large intestine. It is preferable to choose an animal (*e.g.* rabbit or guinea-pig) in which the intestinal coats are thin. The following is the mode of procedure for the chloride of gold method :—A piece of glass tubing about a quarter of an inch in diameter and five or six inches long is taken and one end is drawn out into a cannula, whilst to the other a small piece of indiarubber tube, furnished with a spring clip, is attached. Chloride of gold solution ($\frac{1}{2}$ per cent.) is drawn up into the glass tube so as almost to fill it, and the clip is then closed, to prevent the escape of the fluid. Care should be taken not to suck any of the gold solution into the mouth. A piece of intestine about three inches long is removed from the dead animal, and if not already empty its contents are washed out by a stream of salt solution. The intestine thus emptied and cleaned is ligatured firmly at one end, whilst into the other is tied the cannulated end of the glass tube containing the gold solution. When thus secured the clip is opened and the fluid is allowed to flow into and distend with moderate force the piece of gut, the action of gravity being assisted by gently blowing through the indiarubber tube. As soon as the intestine is filled with the gold solution the clip is again allowed to close, and then, while an assistant holds the glass tube in a vertical position, the operator ligatures the gut just beyond the end of the cannula, which may now be cut away. The piece of intestine, thus filled with the gold solution, is immersed for half an hour in more in the same liquid. It is then placed in a dish of water and cut open longitudinally with scissors, so as to allow the contained fluid to escape, after which the puckered ligatured ends may also be removed. The tissue being hardened by the gold solution, the piece of gut which remains retains its cylindrical shape. It is well to halve it by another longitudinal cut, so that both inner and outer surfaces may be freely exposed to the light. The pieces are now placed, with their outer surfaces uppermost, in a glass vessel of water containing just enough acetic acid to be sour to the taste, and the vessel is covered and allowed to stand in a warm place freely exposed to the sunlight (see p. 120). After two days its colour will be found to have changed to a dark violet. A little methylated spirit may then be added to the fluid ; this serves to aid the reduction of the gold and to prevent the growth of fungi. In another day or two the tissue will be so dark as to appear almost black. A portion is then removed to a glass dish of water and prepared in the following way. In the first place, the glandular mucous membrane is separated from the rest of the intestinal wall, either by tearing it off with forceps or by scraping it away with the end of a blunt scalpel. There now remain the serosa and two muscular layers, together with the submucosa. To the inner surface of the latter the muscularis mucosæ will be still adherent. The separated fragments of the mucous membrane are got rid of by pouring away the water first used and substituting fresh, and then an attempt must be made, by aid of two pairs of forceps, to peel the submucosa off from the inner surface of the muscular coat. Of course if the muscularis mucosæ has been left, that will form a part of the layer which is thus removed. The separation must be done slowly and carefully, so as to get as large a piece as possible intact. When this is accomplished satisfactorily a slide is immersed in the water, and the portion of submucosa so detached is floated on to it, and removed from the water. Its further preparation consists in allowing the excess of water to run off, applying a cover-glass, making sure first of all that the layer is free from folds, and then allowing glycerine to pass under the cover-glass and replace the water as this evaporates.

Returning to the remainder of the piece of intestine, the next process consists in picking away bit by bit with forceps the comparatively thick layer of circular muscular fibres. This is not a difficult proceeding, and when it is finished all that remains is the thin serous coat and the longitudinal muscular layer, to the inner side of which the nervous plexus of Auerbach, the intermuscular plexus, is adherent. No further separation is required, all that is necessary being to float the piece of tissue on to a slide with the (concave) inner surface uppermost. But before applying the cover-glass the preparation is to be examined with a low power, to see that the surface of the serous membrane is free from a finely granular precipitate which is apt to be deposited in the acidulated water. If this is present, the piece must be replaced in the water and the precipitate gently brushed off with a soft camel-hair pencil. The preparation is completed in the same way as that of the submucosa. The latter shows Meissner's plexus, the cords of which are much finer than those of Auerbach's. In both plexuses the nervous cords are stained of a violet colour by the reduction of the gold ; at the points of junction of the nervous cords are groups of small ganglion cells, the nuclei of which are hardly stained at all, and consequently look clear in the midst of the darkly-stained cell-bodies. The distinction between the individual cells is difficult to make out. Branches may perhaps be traced passing from the plexus of Auerbach amongst the muscular fibre-cells : from that of Meissner to the muscularis mucosæ, if this is present, and perhaps also to the small blood-vessels, which are particularly well seen in the preparation of the submucous coat.

This is Cohnheim's method, but Löwit's modification may also be used. In this process the pieces of gut after being removed from the chloride of gold solution are transferred to formic acid solution (1 in 4) and kept in the dark for twenty-four hours or more, until the reduction is complete.

For the chromate of silver method very small pieces of the intestine are taken and treated as described on p. 152. From these sections are cut and are mounted in balsam without a cover-glass.

For the methylene blue method a solution of the dye in 500 parts normal saline is used to distend the piece of gut, which is then tied up and immersed in some of the same solution for about an hour. It is then opened, washed with normal saline, and placed in Bethe's fluid (see p. 160). After the stain is fixed by means of this, the nerves can be investigated either in flat preparations or in sections, mounted in xylol balsam.

Large intestine.—For hardening the tissue and preparing sections of the large intestine the same methods are employed as for the small intestine, so that it is unnecessary to recapitulate them.

The injected large intestine is prepared, like the stomach, by means of vertical and horizontal sections.

The lymphatics are not easy to inject, but present less difficulty than those of the stomach.

CHAPTER XVI

THE LIVER AND PANCREAS

Uninjected liver.—To prepare sections of the liver small pieces are placed in 3 per cent. bichromate of potash solution for ten days, transferred from this to weak spirit, and in twenty-four hours are placed in strong spirit, to complete the process of hardening. Chromic acid (0·2 per cent.), picric acid (saturated), corrosive sublimate, and formol (5 per cent.), may all be used instead of bichromate of potash, but much smaller pieces must be taken, except for formol. They will also be ready to transfer to spirit in a much shorter time. The sections are stained by the usual methods, and mounted in balsam. They should be made in two directions, viz. (1) in a plane near and parallel to one of the surfaces of the liver, and (2) vertical to the surface. Those made in the direction first named will for the most part cut the central or intralobular veins across, those in the second direction may take them along their length; the apparent arrangement of the blood capillaries and liver cells in the individual lobules will differ, both in accordance with this difference of direction and also according as the lobule is cut exactly through its centre or at some part more or less removed from this. Between the lobules are seen the branches of the portal vein, always accompanied by a branch of the bile duct, the columnar epithelium of which is very well seen in these preparations, and by a branch of the hepatic artery. All three are included in a mass of connective tissue, a prolongation of Glisson's capsule, enclosing them in a so-called portal canal.

In this connective tissue cleft-like spaces may generally be seen—two or three in the section of a portal canal—not merely breaks in the connective tissue, but with quite a definite wall. These are the accompanying lymphatics. Other lymphatics accompany the branches of the hepatic veins, but are not so easily seen in the sections, although they can be injected. The branches of the hepatic veins are readily distinguished from those of the portal vein, by the fact that they run unaccompanied by branches of the bile duct and hepatic artery. The blood-capillaries of the lobules look like spaces (filled with round clear bodies,· the altered blood corpuscles) between the rows of cells (in the sections these appear arranged simply in rows); their walls are very thin, and the hepatic cells appear for the most part to come in contact with the wall. But in reality there is a second delicate membrane around many of the capillaries, and between it and the wall of the vessel is a space for the passage of lymph (perivascular lymphatic); it is difficult to make this out, however, in preparations in which the lymphatics are not injected. The round nuclei of the liver cells are deeply stained by the logwood, and the cells themselves slightly. In the thinnest parts of the sections the lines of junction between neighbouring cells can be well made out, and not unfrequently the small capillary passage for the bile which intervenes between the adjacent sides of the cells can, according to the direction in which it runs, be recognised with a very high power either as a horizontal line or as a minute aperture. To obtain the best results the pieces of liver, which are not to be more than an inch or so square and a quarter of an inch thick, should be placed in the bichromate solution quite fresh, from an animal killed only a short time previously.

Injected liver.—The vessels of the liver seldom get completely filled when the rest of the body is injected from the aorta. It is generally necessary to make a special injection of this organ from the portal vein. For this purpose the usual red or blue gelatine injection is used, the apparatus being arranged

THE LIVER 233

as described at p. 179. The operation is conducted as follows :—
The animal (rabbit) having been killed by bleeding,[1] the
thorax is opened ; and the pericardium being torn away, the
heart is raised and two thread ligatures are passed round the
inferior vena cava. One of these is tightened as near the
heart as possible, and then a snip is made in the vein, so as
to allow the blood to escape freely. Next, the abdomen is
opened, and the intestines and stomach being gently drawn
to the left side, the peritoneum at the back of the abdomen is
torn through, and a ligature placed around the vena cava
above the accession of the renal veins. The portal vein is
then found in the fold of peritoneum which connects the
under surface of the liver with the stomach, and a ligature, in
the noose of which the hepatic artery may be included, having
been passed round it near the liver, a snip is made in the
vessel and the injecting cannula is tied in. This cannula is
now filled by means of a pipette with warm salt solution, and
the supply tube (from the injecting bottle) having been completely filled by the injecting fluid to the exclusion of air, in
the same way as previously described (p. 182), is slipped over
the open end, and the injection at once allowed to flow. As
it passes by the portal system of veins through the lobules of
the liver into the hepatic system, it forces whatever blood is
still contained in the bloodvessels of the organ out into the
vena cava, whence it can freely escape into the thorax through
the snip which was there made in the vein. As soon as all
the blood is thus driven out, and only pure injecting fluid
begins to pass, this vein is occluded near the diaphragm by

[1] In injecting the whole body it was recommended to kill the animal by
chloroform. This was for the purpose of having the blood-vessels as much
dilated as possible. When an animal is killed by bleeding, the arteries contract very considerably, and, remaining contracted some little time after
death, offer a considerable resistance at first to the passage of the injection,
and this may tend to spoil the result. In the liver, however, the case is
different, since it is not injected through arteries, but through veins,
which possess little contractility. Any blood which remains in the vessels
does not, so long as it remains fluid, impede the passage of the injection, but
is driven before it.

the second thread. The pressure in the injecting-bottle is then slowly raised, but should not even at the utmost exceed three inches of mercury, for this amount of pressure will cause all the blood-vessels to be quite fully distended, and will effect a very considerable consequent enlargement of the organ ; more might cause rupture and extravasation. After the lapse of a few minutes, to allow of the complete filling of all the blood-vessels, a second ligature is tied round the *portal* vein close to the liver to prevent the return of the still fluid injection, and the cannula is cut out from the portal vein (the pressure in the apparatus having first been removed), and the body put into a cold place so as to permit the gelatine to solidify. The process may be hastened by pouring cold water—iced if possible—over the liver. When the injecting material is entirely set, the organ is removed and cut into small pieces, which are placed in spirit for twenty-four hours or more, or in 3 p.c. bichromate of potash for a few days. When hardened enough, sections may be made (in two directions as with the uninjected organ) and mounted, after passing through alcohol and oil of cloves, in balsam.

During the whole process of injection the greatest care must be taken not to handle the liver more than can possibly be helped, for it is very readily scratched or ruptured and any such accident tends to permit the escape of the fluid injection. This warning applies with equal, if not greater, force to the operation next to be described—that, namely, of filling the bile-ducts.

Bile-ducts.—The bile-ducts are injected with Berlin blue solution, 2 per cent., the mercury apparatus (fig. 56) being used. A simple syringe will, however, answer the purpose. The solution, although fluid in the cold, should nevertheless be employed warm, as it tends to flow more freely. A rabbit is killed by bleeding, the abdomen opened, and the common bile-duct sought for close to the portal vein ; a ligature is passed round the duct and a small piece of card being placed under as a support and to separate it from the accompanying vessels, a snip is made into it, and a glass cannula is inserted, and having been passed along the duct as near to the liver as possible, is tied in. The cystic duct is ligatured to prevent the injection from passing into the gall bladder. In the next place the cannula is filled with warm Berlin blue solution by means of a fine pipette ; the (previously filled) supply tube is attached, the clip on this opened, and the pressure gradually raised. The blue fluid, driving whatever bile there happens to be left in the ducts before it into the lobules, penetrates first into the interlobular bile-ducts, and from these into the outer parts of the lobules, forcing the bile more and more towards the centre ; here of course there is no escape for it, except that a little may pass into the lymphatics and blood-vessels through their walls. Hence it will be understood that the injection can only be made to fill the intercellular biliary passages in the *outer* part of each lobule. The injection should be persevered with for about half an hour ; the bile-duct may then be tied and the injecting apparatus removed, after which the liver is cut out entire, without injuring it in any way, and placed in strong spirit. In three or four hours it is cut into pieces, and the spirit changed, and in less than a week the pieces will be hard enough. The sections may be stained slightly with logwood or eosin.

Method of Golgi.—The bile-ducts are extremely well shown by this method, which must be applied to small pieces of the liver in the manner described on p. 152. Nerve terminations may also be seen in this preparation, and the reticular connective tissue of the organ is also sometimes very well displayed.

Lymphatics of the liver.—The lymphatics of the liver are injected through a fine cannula stuck obliquely into the superficial part of the organ immediately beneath the capsule. Either solution of Berlin blue or alkanet may be used. The part should be quite fresh. If the injection be persisted in for a long while, the fluid may flow out both by the lymphatics accompanying the portal vein and those accompanying the hepatic veins (Ludwig and Fleischl) Very frequently, however, the injecting fluid finds its way into the blood-system instead of the lymphatics. The injection of the lymphatics may be accomplished in another manner, viz. by seeking the lymphatics which accompany the hepatic veins at the back of the liver, and tying a cannula into one of them. After a time the fluid will be found to pass out by the vessels which accompany the portal vein.

Hepatic cells.—In addition to what may be learnt from sections of the organ, teased-out preparations afford much useful information, both of the characters of the liver cells and of the connective tissue of the lobules. For this purpose small portions of the perfectly fresh and warm liver are broken up in serum or salt solution, and other portions are macerated for a day or two in one-third alcohol, and subsequently teased out in water, and stained with dilute hæmateïn.

Demonstration of glycogen in the hepatic cells.—This is most readily shown microscopically in the liver of a rabbit which has been fed a few hours previously with a meal of carrots. The animal having been killed rapidly, by bleeding or otherwise, small pieces of the liver are thrown into strong spirit, and left in this until sufficiently hardened. Thin sections are then cut by the paraffin method, and are treated, after being passed through xylol and absolute alcohol, with a solution of iodine in iodide of potassium, which stains the cells which contain glycogen of a reddish brown colour. The sections may either be mounted in diluted glycerine or in acetate of potash solution. Alcohol, xylol, chloroform, &c. rapidly extract the colour.

Demonstration of iron in the hepatic cells.—This can be effected by the employment of Macallum's method (p. 24). It will be found necessary to treat with hydrochloric acid prior to hæmatoxylin in order to exhibit the iron here.

The **pancreas** is prepared in the same manner as the salivary glands, to the description of which the student is referred. The zymogen granules are well shown in sections double-stained with carmalum or hæmalum and picric acid ; and these methods of staining also bring out well the peculiar tissue of the 'islands.'

CHAPTER XVII

THE DUCTLESS GLANDS

Ductless glands of the larynx and trachea.—The thyroid and **thymus** are studied chiefly by means of sections, for facilitating the preparation of which the glands are hardened in alcohol alone, or in sublimate, picric acid, or formol, followed by alcohol. Unless from a small animal they should not be put entire into the fluid, but cut into thin pieces, so that the preservative fluid may penetrate rapidly. They are best stained by hæmateïn, which colours, besides the nuclei of the cells, the so-called 'colloid' which is met with in the vesicles of the thyroid.

The concentric corpuscles of Hassall which are met with in the thymus can be seen in sections of that organ, but may also be studied isolated in preparations of the fresh gland teased out in salt solution. They are stained yellow by picric acid. The **pituitary body** and **pineal gland** are prepared in the same way as the thymus and thyroid.

Sections of lymphatic glands.—These are chiefly studied by means of sections. They are hardened in strong spirit, into which they are put immediately after removal from the animal; in two or three days they are sufficiently firm to cut, but improve if left longer. Or they may first be injected interstitially with dilute chromic acid (0·2 per cent,) and placed in a quantity of the same fluid for a few days, then transferred to 50 per cent. spirit, which is to be changed in two or three days to strong alcohol. They may either be cut from paraffin or soaked in gum and cut by the freezing method.

The sections, which *must* be very thin indeed, and should include both cortical and medullary substance, may be mounted in glycerine, without staining, or they may be stained with hæmateïn and mounted in xylol balsam. If the lymph-paths appear filled up with lymph corpuscles, so that the retiform tissue which traverses them is not well seen, but the whole section appears more or less uniform in structure, these corpuscles may be in great measure removed by vigorously shaking up the sections with water in a test-tube, or by gently brushing them under spirit with a soft camel-hair pencil. Unfortunately both these methods tend to break up the sections, and indeed it is not necessary to employ them if the sections are made sufficiently thin.

The glands may also be stained in bulk, a thin piece being placed in carmalum or some other good bulk-stain for 24 hours or more; then either soaked in gum and cut frozen or embedded in paraffin, and the sections fixed on the slide and mounted in balsam in the usual way.

Reticular tissue of lymphatic glands.—This may be made manifest in various ways. One of the simplest is to stain a section with acid fuschin dissolved in alcohol, then clear with oil of cloves and mount in xylol balsam. Or a piece of fresh gland may be frozen and sections cut, and dipped for a minute in a 1 per cent. solution of caustic potash, then thoroughly washed, stained with acid fuschin, and mounted as before. The methyl-blue and eosin stain recommended on p. 23 will also be found valuable for differentiating the fine retiform tissue, since this, like other connective tissue, becomes stained of an intense blue by that combination.

The retiform tissue may also be well seen in preparations made by Golgi's method (p. 152). The lymphatic glands of the dog may be recommended for demonstrating the structure of these organs, especially for showing the trabeculæ and lymph-sinuses. In some animals the trabeculæ are far less developed, and the gland is little but a mass of lymphoid tissue intersected by lymph-channels.

THE SPLEEN

The uninjected spleen.—The spleen is hardened in the same manner as the liver. The sections are to be stained with hæmateïn and eosin or carmalum and picric acid, and mounted by the ordinary modes of procedure.

In these preparations the Malpighian corpuscles (or nodules of lymphoid tissue) are very strongly coloured by hæmateïn ; the trabeculæ which traverse the pulp, which are largely composed of plain muscular tissue, are stained yellow by the picric acid ; the substance of the pulp is but slightly coloured by the logwood, only the cell-nuclei, and, to a much less extent, the network of the retiform tissue being stained. The prevailing colour of the pulp is yellowish, owing to the blood, which at the time of death remained in the interstices of the tissue becoming stained by the picric. Here and there a small mass of coarsely-granular reddish-yellow pigment may be detected, lodged in one of the corpuscles of the spleen pulp. With a high power the manner in which the veins open out of the interstices of the pulp may be made out.

Irrigated spleen.—By another mode of preparing the spleen all the blood is first washed out by a stream of salt solution, injected through the splenic artery, and the salt solution is followed by a stream of 3 p.c. bichromate or picric acid. This is made to distend the organ somewhat, the distension being maintained by ligaturing the vessels near the hilum. The organ is now placed entire in a quantity of the solution, and only cut into pieces after forty-eight hours. By thus removing the blood-corpuscles the retiform tissue of the pulp is better seen.

Injected spleen.—The spleen may have been injected in the animal which was injected entire ; if this is not the case, a special injection is to be made from the splenic artery. When successfully accomplished, the vessels are as usual ligatured to prevent the escape of the injection. When the gelatine has set the organ is cut up and pieces placed either in 3 per cent. bichromate of potash or in 75 per cent. spirit. The sections will show what at first sight look like accidental extravasations; patches, namely, of injection distributed all over the organs, with the exception of large round white masses, here and there, pervaded by a few capillaries. The white masses are sections of the Malpighian corpuscles, and the part permeated by the injection is of course the pulp, into which the arterial capillaries freely open.

Splenic cells.—To obtain specimens of the spleen substance, which will show in a separated condition the cellular elements which it contains, and of which it is composed, a small portion of the fresh organ may be teased out with needles in a little salt solution or serum. But it will be found that so much blood is incorporated with the spleen substance (it forms, in fact, most of the soft matter which can be expressed from the fresh section) that the view of the other parts is obscured by innumerable red blood corpuscles. Hence before teasing a piece it should be placed for forty-eight hours in one-third or one-fourth alcohol. This decolourises the red corpuscles whilst preserving the character of, and at the same time macerating somewhat, the proper substance of the spleen, so that the cells are now readily separated and seen. By far the greater number are lymph corpuscles from the lymphoid tissue of the Malpighian nodules and of the arterial adventitia. But, besides these, other cells (splenic cells) are met with; larger, rounded or flattened, and some of them containing pigment granules as already intimated. They may be found either entirely isolated or adhering to the network of the retiform tissue. Their nuclei, as well as those of the lymphoid cells, are well brought out if a little hæmateïn solution is permitted to pass under the cover-glass. In the fresh preparations, not treated by bichromate of potash or any other reagent, but made in serum, some of the cells may perhaps be found containing red blood corpuscles in their interior, and transitions from these to those containing pigment are met with.

Reticulum.—The reticulum of the spleen is best exhibited in preparations made by the silver chromate method of Golgi.

THE SUPRARENAL CAPSULES

Sections.—To prepare the suprarenal capsule it is separated from the surrounding fat, divided into two or three pieces by transverse cuts, and placed in 3 per cent. bichromate of potash solution for fourteen days, when the hardening may be completed in spirit in the usual manner. Hardening the organ in corrosive sublimate or in formol or in picric acid followed by spirit also gives good results. The mode of preparing the sections, which should include both cortical and medullary substance, calls for no special description.

Sections of the injected organ should also be prepared.

Teased preparations.—In a teased-out preparation of the fresh organ the cellular elements of the cortical and medullary substance may respectively be studied, and the effect of a solution of bichromate of potash in colouring the medullary cells brown may be observed.

CHAPTER XVIII

THE KIDNEY

The uninjected kidney.—The kidney is hardened in the same way as the liver and spleen; the yellow chromates of potash and of ammonia may also be employed (in 5 per cent. solution). The piece or pieces that are taken should include both cortical and medullary parts; but at the same time should not be thicker than from an eighth to a quarter of an inch, otherwise the preservative fluid will not penetrate rapidly enough to the deeper parts. They are to remain in the bichromate solution, if this is employed, for two or three weeks; are then placed in weak spirit, and in twenty-four hours transferred to strong spirit. In three or four days more the pieces are firm enough to cut. Or, after being hardened by the bichromate solution the pieces may be simply soaked in gum and cut frozen, without passing through spirit at all. Sections as thin as possible are to be made in a plane vertical to the surface of the organ, and large enough to include both cortical substance and Malpighian pyramid. One such section is simply mounted in glycerine; others are stained, and after treatment by the customary processes mounted in balsam. These sections will show—in the cortical substance the Malpighian corpuscles, the convoluted tubules variously cut, and the prolongations of the straight tubules of the medullary substance and of the tubules of Henle; in the pyramidal part the two last-named tubules, and the collecting and excretory tubules, seen longitudinally, with a large number of blood-vessels running parallel to and between them.

Transverse (tangential) sections also should be obtained both from the cortex and from the medullary substance as well as sections across a papilla.

The injected kidney.—The blood-vessels of the kidney will very probably be filled in injecting an animal entire ; but, if this should not have been the case, it is not difficult to make a special injection of the separated organ from the renal artery. The red gelatine injection may be used, and the kidney is kept warm, and the injection maintained for a considerable time, in order that the vessels of the glomeruli and the network of capillaries in the cortical substance supplied by their efferent vessels may be completely filled. The organ is then set aside in a cool place (surrounded by ice, if possible), and, when the gelatine is completely set, is cut into three or four pieces and hardened, as usual, with alcohol or bichromate of potash. The sections, which need not be very thin, but should be quite even and comprise the whole thickness of the organ, are to be mounted, unstained, in balsam.

Uriniferous tubules.—The uriniferous tubules may be injected from the ureters for a considerable part of their length, simultaneously, if it be desired, with the above-described injection of the blood-vessels, by a solution of Berlin blue. But even if well filled they are too densely arranged to render it possible to trace individual tubules along their whole extent, except in very young animals. This may be better accomplished by making teased preparations of the kidneys of small animals, which have undergone some process of preparation, having for its object the solution or softening of the intertubular substance. Several such processes have been proposed, but none yield entirely satisfactory results. The best, perhaps, consists in digesting tolerably thick slices of a small kidney in a mixture of 4 parts of spirit and 1 of hydrochloric acid, kept boiling for 3 or 4 hours. The boiling is performed in a flask fitted with a condenser, in which the vapour which is driven off by the boiling becomes condensed and flows down again into the flask. The slices are then placed in water, and, after lying in this for a few days, minute shreds, comprising the whole depth from external surface to papillæ, are split off with needles, placed on a slide, and unravelled as much as possible by aid of the dissecting microscope. The preparation is covered with a specially large piece of covering-glass (a hair being first added to avert the pressure of the glass on the soft tubules), and stained by drawing picric acid solution under the cover-glass. This soon colours the tissue intensely yellow; glycerine may then be allowed to pass in at the border in order to complete the preparation. Some of the tubules will be found isolated for a considerable part of their length, and the passage of the convoluted tubules into the looped tubes of Henle may especially be well seen. The epithelium of the tubules is for the most part well preserved.

Examination of the fresh kidney.—When by these various means sufficient acquaintance has been gained with the various tubules and their contents *in situ*, the examination of the fresh tissue in serum may be attempted. With this object small snips are to be made from different parts of a freshly-cut surface with a pair of curved scissors and teased out in a drop of serum, with the aid of the dissecting microscope, so as to separate as many of the tubules as possible. In doing this much of the epithelium will become detached, and the characters of the individual cells in the fresh condition may be studied.

The outlines of the cells of the capsules of Bowman and of the tubules may be shown by nitrate of silver. For this purpose a fresh kidney is sliced in half by a single cut with a sharp razor in the direction of the tubules. One of the halves is thoroughly washed with distilled water, and solution of nitrate of silver ($\frac{1}{2}$ per cent.) is poured over the cut surface. After a minute and a half the silver solution is rinsed off with distilled water, and the piece of kidney is placed in water, with the silvered surface exposed to the sunlight. When brown it may be removed from the light, and placed in spirit for twenty-four hours; one or two sections are then made from the brown surface, clarified in oil of cloves, and mounted in balsam.

THE URETERS

Sections.—The ureters are prepared in the same way as the intestine—by moderately distending an excised portion

THE URINARY BLADDER 245

with some sort of hardening solution, and placing the piece thus distended in a beaker containing some of the same mixture. After a few hours the tube is slit open, and transferred to spirit after two or three days. The sections are to be made across the length of the tube, and stained and mounted in the ordinary manner.

Epithelium of ureter.—To study the separated epithelial cells a piece as fresh as possible is cut open, pinned out on a cork with the inner surface uppermost, and immersed in one-third alcohol or $\frac{1}{8}$ per cent. bichromate of potash solution for from twenty-four to forty-eight hours. Some of the epithelium is then scraped off with a spear-shaped needle or the end of a scalpel, and is broken up in a drop of water. After the addition of a piece of hair to the fluid the cover-glass may be applied, and the preparation examined with a high power. If it prove successful, with many of the epithelial cells fully separated, it may be permanently preserved. With this object the cells should first be stained, by allowing weak hæmalum solution to run under the edge of the cover-glass. The hæmalum is to be followed by a drop of glycerine applied at the same edge; and, when the glycerine has become diffused underneath, all that is necessary is to cement the cover-glass.

THE BLADDER

The **urinary bladder**, both for sections and teased-out preparations, is prepared by exactly the same methods as the ureters. To distend it a glass cannula, connected by an indiarubber tube with a bottle containing the hardening fluid, is tied into the urethra. The organ must not be over-distended, but only moderately filled. Any urine which it may contain should first be allowed to run out through the cannula.

In dealing with the human bladder or the bladders of large animals it is more convenient to cut out one or two pieces from different parts and to pin them on a cork or cake of wax for immersion in the hardening fluid.

The musculature of the bladder in the frog has already been studied (p. 135). Such a preparation can also be made of the bladder of a small mammal, and will serve to show the lymphatics as well. The nerves may be shown by any of the special methods detailed in the chapter on nervous tissue.

CHAPTER XIX

THE GENERATIVE ORGANS

Erectile tissue.—Those parts which contain erectile tissue will be best studied after having been injected. Their bloodvessels and sinuses may have been filled in the animal which was injected entire from the root of the aorta; but if not, a special injection from the lower end of the abdominal aorta is to be made, the arteries supplying the lower limbs being first tied to prevent waste of the injection. The hardening of the parts in spirit must be effected very gradually (the spirit being daily made stronger), since otherwise the gelatine shrinks away from the walls of the venous sinuses, and the preparation becomes in great measure spoiled. This is, however, obviated by using bichromate of potash. The sections should some of them be mounted, unstained, in balsam, others after being lightly stained with hæmateïn, so that the plain muscular and fibrous tissue, and also, in sections including the urethra, the epithelium of that tube may be exhibited as well as the vessels.

Parts which have not been injected are hardened in 3 per cent. bichromate of potash solution, 0·2 per cent. chromic acid or saturated solution of picric acid, followed by spirit in either case.

Prostate and seminal vesicles.—The glandular organs, such as the prostate and vesiculæ seminales, are prepared in the same way or with spirit only.

Scrotum, labia, &c.—The scrotum, and labia, and the vagina are prepared in the same way as the skin (see p. 199).

Uterus.—The human uterus is best hardened in 3 per cent. bichromate of potash; its cavity should be freely laid open. Sections may be cut from it by the freezing method. In animals (the rabbit, for instance), where it is more membranous, the uterus and the upper part of the vagina may be prepared together by distending them with picric or chromic acid solution through a cannula tied into the lower part of the vagina. The vagina is then tied up, and the organs are cut out and placed in a quantity of the solution; in three or four hours they are laid open and the fluid renewed, and in a day or two are ready to be put into spirit. The sections are stained and mounted in the usual way.

Section of ovary.—The ovaries are prepared by placing them—with as little handling as possible, so as to avoid rubbing off the columnar epithelium which covers the surface—in 0·2 per cent. chromic acid or in picric acid or in formol (whole if taken from a small animal, such as a rabbit or cat; cut into two or three pieces if from a larger one). In most of the lower animals they must be sought much higher in the abdomen than in the human female; in the rabbit they occur as small elongated bodies, dotted all over with little projections (the Graafian follicles), and situated just below the kidneys. They are left in the chromic solution for seven days, and then placed in spirit, which must be changed frequently, and in a few days more are ready for cutting. The sections are to be stained in hæmalum, carmalum, or picro-carmine solution; hæmalum sometimes colours very deeply the coagulated fluid in the Graafian follicles, so that the epithelial contents are obscured.

The ovum.—The ripe mammalian ovum, although it can be seen within the larger Graafian follicles in the sections of the hardened organ, forms a much more beautiful object when obtained isolated from the ovary of a recently-killed animal. A full-grown doe rabbit, not pregnant, is to be sacrificed for this purpose. One of the ovaries having been removed, it is held firmly between the finger and thumb over a clean glass slide, in such a position that the largest and most prominent Graafian follicle is almost in contact with the middle of the slide. The follicle is then pricked with a sharp-pointed scalpel, so as to cause the fluid contents of the follicle to spirt out, carrying with them the ovum, surrounded by a number of the epithelium cells. The ovum is rather too small to be detected with the naked eye, but its presence may be suspected if, on glancing at the slide in such a manner that the light is reflected from the surface of the fluid to the eye, a slight prominence is observed on the otherwise flat surface. Its presence here is confirmed by examination with a low power, and it may then be carefully observed with the ordinary high power. It is better, if possible, not to apply a cover-glass, for the zona pellucida is apt to become broken; and, moreover, even slight pressure spoils in great measure the natural appearance of the object. But if the objective becomes dimmed by its proximity to the fluid, or if it is desired to employ an immersion, then a thin cover-glass must be used, and to protect the ovum from pressure a narrow slip of thick paper (an ordinary hair is not thick enough) is to be put on either side before the cover-glass is applied. If the fluid which accompanies the ovum from the Graafian follicle is not in sufficient quantity, a drop of aqueous humour may be added to it. It is not possible to preserve this preparation very satisfactorily.

Section of uninjected testis.—The testis is best hardened in formol or absolute alcohol. If from a large animal it should be cut into in two or three places before being plunged in the fluid. It varies in consistence, being firmest in those animals (cat, pig) in which the peculiar granular polyhedral cells, which accompany and surround the blood-vessels, are most numerous. Thin pieces may be stained in bulk with dilute hæmalum or carmalum. Sections of both the body of the testis and the epididymis are to be made.

Lymphatics of testis.—In the sections made as above there will be observed in the interstices between the seminiferous tubules large cleft-like spaces, looking almost like accidental clefts in the loose connective tissue uniting the tubules. They are in reality, however, the lacunar commencements of the lymphatics. To show this the following simple experiment may be performed :—In a recently-killed dog the fine cannula of a hypodermic syringe, filled with Berlin blue solution, is stuck through the scrotum into the middle of the substance of one of the testes, and the solution is slowly, and without exerting any considerable pressure, injected into the organ. If the abdomen is opened the blue fluid will soon be seen passing along the lymphatics which run in the spermatic cord, and from these into those of the back of the pelvis and abdomen, at length reaching the thoracic duct. If now the testis is removed and hardened in spirit, and sections are made of the hardened organ, it will be found that the intertubular spaces are occupied by the blue substance, and, since they are proved by the injection to be in free communication with the lymphatics which leave the organ, the spaces are to be looked upon as giving origin to the lymphatics.

Isolation of the seminiferous tubules and elements.—For obtaining the tubules isolated for a considerable length, pieces of the testis (preferably human) are placed for a day or two in hydrochloric acid, diluted with one-third its volume of water, and maintained at 30° C. They are then allowed to lie in water until the tubules can be readily separated with needles. Teased-out preparations of fragments of testicle which have been macerated in one-third alcohol for two or three days may also be made. Fragments of the fresh testis-substance should also be teased in saline, to exhibit the form, stages of development, and movements of the spermatozoa. For the object last named the preparation should be examined on the warm stage.

Epithelioid cells of seminiferous tubules.—To exhibit the fact that the apparently structureless basement membrane of the seminiferous tubules is in reality composed of layers of flattened epithelioid cells, a portion of the testicular substance is partially unravelled in distilled water, and some of the

THE MAMMARY GLANDS 251

tubules which are thus isolated are dipped into nitrate of silver solution for a minute, and, after being again rinsed in water, are mounted in glycerine and exposed to the light; the lines of junction between the flattened cells are by this means made evident.

Tunica vaginalis.—The tunica vaginalis is to be prepared in the same way as the other serous membranes (with nitrate of silver), partly unbrushed to show the epithelioid covering, and partly brushed for the sake of exhibiting the parts beneath. For the preparation of the visceral part, the process is similar to that adopted for the pericardium covering the surface of the heart (p. 205), and need not here be more particularly described.

The mammary glands.—Small pieces of these organs are hardened by being placed in 0·2 chromic acid solution for a week or ten days, or in picric acid for a day or two, or in formol or corrosive sublimate, subsequently transferring them to spirit. Absolute alcohol may also be used at once, and produces the desired result more rapidly; the sections stain more readily than if chromic or picric acid is used. The appearances presented by the cells of the alveoli vary considerably according to the state of functional activity of the gland. It will also be found that the human mammary gland is much less compact, its glandular substance being more scattered in the connective tissue of the organ than is the case in most animals.

CHAPTER XX

THE CENTRAL NERVOUS SYSTEM; THE BRAIN AND SPINAL CORD

Preparation by silver chromate.—Some experience has already been obtained of the methods which are employed for studying the cellular elements of the central nervous system (pp. 150-153). The silver chromate method of Golgi is to be employed as there directed for studying portions of the cerebral cortex, of the cerebellar cortex, of the medulla oblongata, and of the spinal cord. In all cases it will be of advantage to employ fœtal or young animals, but this is not necessary, for very good preparations, especially of cerebral cortex, are yielded by adult organs treated by this method. The spinal cord was the part then under investigation, but the nerve-cells found in the grey matter of the cerebrum and of the cerebellum may be observed in the same manner with equally satisfactory results. Without delaying, then, to repeat the directions there given, we may pass on to describe the best general methods of preparing sections of the parts in question.

Preparation with bichromate of ammonia.—To harden any part of the central nervous system for histological purposes the most generally useful reagents are the bichromates of potash and of ammonia (2 or 3 per cent. solution). These will in three or four weeks render the tissue sufficiently firm for obtaining thin sections, which may be made, after a preliminary soaking in gum, by the freezing method. For celloidin-embedding it is necessary to complete the hardening first with weak and then with strong spirit. The pieces to be hardened should not be too large, or at least not too thick, but the bichromate solution has considerable power of penetration, and the whole length of the spinal cord of any of the smaller quadrupeds, and even that of man, may be hardened in it intact if put in perfectly fresh. It is always better, however, to cut it into short lengths. Ten per cent. of formol solution has of late come into use in place of bichromate of potash. It penetrates even more readily, and hardens much more rapidly. To stain by Nissl's method (see next page) alcohol only may be employed.

In this way a piece of the cerebellum, two or three small pieces from different parts of the convoluted surface of the cerebrum, the medulla oblongata, and pieces of the spinal cord from the middle of the three regions (cervical, dorsal, and lumbar) are to be prepared.

Preparation of sections of the central nervous system.—Although it is possible to cut small pieces of the central nervous system sufficiently thin either by the freezing method or by an inclined plane microtome, with the knife wetted with spirit, for larger sections and when it is very necessary to keep the neighbouring parts (nerve-roots and the like) in position, the collodion method must be used (see p. 29). For staining by Nissl's method, it is desirable to imbed in paraffin.

Staining of the sections. Weigert-Pal method.—Sections which have been prepared by the bichromate method may be stained in the ordinary way with carminate of soda (5 p.c.), carmalum, or aniline blue-black. Acid fuschin has also been recommended by Weigert as a reliable stain. A saturated water solution is used, and the sections are partly decolourised by absolute alcohol rendered very faintly alkaline by caustic potash, and are then washed with water and alcohol, and mounted in Canada balsam. But for exhibiting the medullated fibres, and the tracts which they form, the following modification of a method originally devised by Weigert, and known by his name, will be found the best :—The sections, whether cut by the freezing method or by the collodion process, are placed first in water for a few minutes ; then for a few hours in Marchi's solution, which is a mixture of Müller's fluid (2 parts) with 1 per cent. osmic acid (1 part). They are then washed in water and transferred to acid hæmatoxylin (p. 19), in which they are left overnight ; by which time they are completely black. They are next thoroughly washed with tap water and put first into 0·25 per cent. solution of permanganate of potash for five minutes, then rinsed in water and transferred to Pal's solution (sodic sulphite 1 g. ; oxalic acid 1 g. ; distilled water 200 c.c.). This bleaches the grey matter, leaving the medullary sheath of all nerve-fibres black. The process of differentiation usually takes but a few minutes, but if they are not properly differentiated in half an hour they are replaced, after washing, in the permanganate solution for a few minutes and then again placed in Pal's solution. Finally they are well washed with tap water, and transferred through alcohol of increasing strength and oil of bergamot into xylol balsam. This modification has many advantages over the original methods of Weigert and Pal.

The sections may before mounting be stained with eosin to show the nerve-cells.

Nissl's method.—The tissue is hardened in alcohol (96 p.c. or absolute), which is changed after the first day. It is left in alcohol from two to six days, then simply embedded, not soaked, in paraffin. The sections are stained, after removing the paraffin, by immersion for a few minutes in a 0·5 p.c. aqueous solution of methylene blue, which has been just previously heated to about 70° C. for half to one minute. They are then differentiated by alcohol or aniline-oil alcohol (1 to 5), passed through origanum oil, and mounted in xylol balsam. The nerve-cells and their processes are beautifully stained, the methyene blue being retained especially by the cell-granules. A double staining may be got by using alcoholic solution of

NERVOUS SYSTEM 255

eosin of orange G, or of acid fuschin, as a differentiating reagent.

Weigert's method for showing neuroglia-cells.—After hardening in 2 per cent. bichromate of potash, followed by alcohol, sections are cut by the celloidin method, and first stained slightly with carmine. They are next stained with gentian-violet as follows :—A water solution of the dye is made by adding a few drops of a saturated alcohol solution to aniline water (see p. 21). After staining, the sections are decolourised in iodine solution (1 per cent. dissolved in 2 per cent. solution of potassic iodide). After two or three minutes they are transferred to a mixture of aniline oil (2 parts) and xylol (1 part). When sufficiently decolourised they are well washed with xylol and mounted in xylol balsam.

Tracts of conduction in the central nervous system.—To study the tracts of conduction in the central nervous system two methods may be employed. One is to prepare by the Weigert-Pal process sections from the several parts in the fœtus and new-born infant. Those tracts in which the medullary sheath of the nerve-fibres is still incomplete appear in such sections unstained, whereas those in which it is complete are stained black : in this way many of the principal tracts of conduction (*e.g.* the pyramidal, which develop late) are differentiated from the rest of the white matter. The other method consists in producing degenerations of certain tracts and staining the degenerated nerve fibres. By the Wallerian law of degeneration, any nerve-fibre which is separated from its nerve-cell undergoes a process of degeneration which is characterised by its medullary sheath breaking up into droplets of myelin. These droplets of myelin are much more susceptible to osmic stain than normal medullary substance, and advantage is taken of this fact in the method employed to show the tracts.

The first thing is to produce the required lesion, such as a complete transverse section or a hemisection of the spinal cord, a removal of a portion of cerebral cortex, and the like.

The operation is done aseptically and the external wound should heal at once. The animal is kept alive for two or three weeks. It is then killed, and the organs to be investigated are placed in 2 per cent. bichromate solution or Müller's fluid for a few days; thinnish slices are then cut from different parts and laid on cotton-wool in a considerable quantity of Marchi's solution (each one in a separate vessel), which should be changed in a day or two if it lose the odour of osmic acid. They are left in the Marchi fluid for a week, and are then washed with water and placed in spirit. Sections are cut by the collodion method, and are not further stained, but are simply passed through alcohol and cedarwood oil to be mounted in balsam. The degenerated tracts stand out, even with the naked eye, as black patches upon the generally colourless ground of the section.

It has not hitherto been possible to produce this Marchi staining of degenerated nerve-fibres in individual sections, but only in thin pieces of the tissue. Hamilton has devised a method whereby the staining can be effected in the sections. Instead of taking fresh bichromate of potash, he uses bichromate solution which has been already used for hardening pieces of brain, adds 1 c.c. of 1 per cent. osmic acid to 200 c.c. of this solution, and keeps the sections twenty-four hours in the mixture at about 38° C. They are then washed and transferred for a short time to the following solution:

Pyrogallic acid	1 grm.
75 per cent. alcohol	50 cc.
Glycerine	50 cc.
Distilled water	300 cc.

then rinsed in water, placed for 3 days in 0·25 per cent. permanganate of potash, washed with solution of sulphurous acid, and mounted in balsam in the usual way.

257

CHAPTER XXI

THE ORGANS OF THE SENSES—THE EYE

THE study of the eye should be made as much as possible from that of the human subject, for there are slight differences in the structure of some of the parts in man and animals, and, moreover, it is on the whole easier to demonstrate the structures in the human eye. On the other hand, there is no organ which it is so absolutely essential to obtain in a perfectly fresh condition. For this and other reasons it is scarcely possible to get material from the post-mortem room, and the only opportunities that usually present themselves occur when an eye is removed on account of some injury or disease which is confined to a particular part; the other intact portions may in such cases be available for histological purposes. In rare instances an entire healthy eye has to be removed, and the following would be perhaps the best way to deal with the excised organ with the view of making the best use of it :—As soon after removal as possible separate the eye, by an oblique cut with a very sharp knife or razor, into two halves, an anterior and a posterior ; the cut to start from just behind the attachment of the iris anteriorly and superiorly, and pass downwards and backwards towards the posterior part of the organ, coming out a little below the yellow spot and optic nerve. Then put the posterior part, after allowing the vitreous humour to fall away from the retina, into a mixture of three parts of Müller's fluid with 1 part of 1 per cent. solution of osmic acid ; the anterior part may be put into Müller's fluid alone. The cornea is to be cut through at one place with a

s

sharp scalpel, so that the preservative fluid may get freely into the anterior chamber.

The piece in Müller and osmic is left there for two to five days or more : each day a small fragment of retina is to be removed and prepared with silver nitrate by the method of Golgi (p. 152). What remains of the preparation is after five days placed in water for two hours, and finally transferred to a mixture of equal parts of glycerine, alcohol, and water ; in this it is to remain for a week or more until wanted.

The other piece is to lie in Müller's fluid for three weeks, changing the fluid once or twice ; it is then placed in weak spirit for a day or two, and finally preserved in strong spirit. Formol (10 per cent.) may be used in place of Müller's fluid ; it will be found much more rapid in its action. Of the lower animals, the eyes of the pig serve best for exhibiting the minute structure, especially of the retina. In this animal the eye corresponds more closely in point of size and approaches more nearly in structure to the human eye than that of the ox or sheep, the other animals the eyes of which are usually readily procurable.

The eyelids.—These are studied by making sections of the hardened lid across its long axis and vertically to its surfaces. The lid may be obtained from a still-born child, preferably one the blood-vessels of which have been injected. It is to be hardened in spirit and embedded, and the sections—which present no unusual difficulty—stained and mounted in the usual way. In this way almost all the parts are well displayed : the skin with its epidermis on the outer side, and with a few small hairs and sweat-glands seen here and there ; the mucous membrane (conjunctiva) on the inner side ; the Meibomian glands cut along the length of their wide, straight duct, with their round saccules lined with epithelium cells (of a whitish glistening appearance, due to the fatty secretion they contain, and which also fills the duct) ; the eyelashes with their large hair-follicles and sebaceous glands ; the cut ends of the very small muscular fibres of the orbicularis arranged in groups, separated by connective tissue; and the general connective tissue which serves to unite all the parts together and, becoming denser towards the inner surface, forms the so-called 'tarsal cartilage,' long described as composed of fibro-cartilage, in reality containing no cartilage-cells.

The lachrymal gland is prepared in the same way as the salivary glands.

The substance of the sclerotic.—The sclerotic is studied by means of sections made from an eye that has been hardened in 3 per cent. bichromate of potash. They are made frozen, after the tissue has been soaked in gum. Covering the outer surface of the globe is a loose connective tissue membrane, the capsule of Tenon, composed of two apposed layers, lined by epithelioid cells, which bound a lymph-space, and covering the inner surface of the sclerotic is another delicate lamella of loose connective tissue, closely adherent to the fibrous substance of the coat, and of a brown appearance, due to the presence of pigment. This layer (the lamina fusca) is also bounded internally by a layer of epithelioid cells, and is separated from a similar lamella on the outer surface of the choroid coat (the lamina suprachoroidea) by another lymph-space, traversed here and there by the vessels and nerves as they pass obliquely across it from the sclerotic to the choroid.

Capsule of Tenon.—To exhibit the epithelioid cells of the capsule of Tenon a fresh eye is taken, and the adhering orbital fat, and everything but the insertions of the eye-muscles, removed from the globe, except the above-mentioned loose membrane. The eye thus cleared is rinsed in distilled water and a few drops of nitrate of silver solution are poured over the posterior part. After three minutes the silver is rinsed off again by a stream of distilled water, and the eye is placed in water in the sunlight. When sufficiently stained it is removed from the window, fastened under water to a loaded cork by a long pin passed through the cornea, and a piece of the capsule of Tenon is dissected off the globe, floated flat on to a slide, and removed from the fluid. After the excess of water has been got rid of, the piece is covered in glycerine and examined for the epithelioid markings. Or the preparation may be made by hardening in spirit and preparing tangential sections of the silvered surface.

Lamina fusca.—The epithelioid layer lining the lamina fusca is also prepared by nitrate of silver. A piece of the sclerotic is dissected off from a fresh eye; the convex outer surface of the piece is then pressed in and made concave, the previously concave inner surface being made the convex one, and the piece is first dipped in distilled water, then placed for two minutes in silver solution, then, after being again rinsed in water, transferred to spirit and placed in the light, with the inner surface or lamina fusca uppermost. After a few minutes, by which time in bright diffused daylight the silver will probably be reduced, although owing to the natural brown colour this cannot well be seen, it is removed, and when hard enough for the purpose, sections are made from the brown surface, placed in water, and mounted in balsam. The pigment-cells usually obscure the silver markings to a certain extent. This inconvenience can be obviated by using the eye of an albino rabbit. Here, moreover, the sclerotic is not too thick to admit of the piece being mounted entire, two or three radial slits being made in it if necessary, with the object of causing it to lie flat on the slide.

Besides this preparation of its epithelioid layer, the lamina fusca may itself be displayed in an eye, or portion of an eye, that has been prepared with bichromate. To obtain it, a small piece of the sclerotic is pinned to a cork or wax-cake under weak spirit (equal parts of water and spirit); and the lamina fusca is dissected off from its inner surface, and floated on to a slide, the spirit being then allowed to evaporate, so as to leave the delicate membrane moistened only with water. The preparation may now be covered, and glycerine added at the edge of the cover-glass.

Sections of cornea.—The several layers of which the cornea is composed, and their relative thickness, should first be studied in sections made vertically to its surfaces. For this purpose the anterior part of an eye (pig's or ox's if a human eye is not procurable) is placed in 3 per cent. solution of bichromate of potash or Müller's fluid for two or three weeks, or in 10 p.c. formol for a few days. It is as well to remove the lens so that the solution passes freely to the posterior surface of the cornea. After the time specified the tissue is put into weak spirit for twenty-four hours, and then transferred to strong spirit. In two or three days more it will be ready for making sections. For this purpose a piece of the cornea is cut out and embedded in paraffin. Very thin sections vertical to the surface are to be made, and stained and mounted in the usual way in balsam. Sections can also be easily made by the freezing method, and after being stained with hæmalum or carmalum mounted in balsam. In this process a source of difficulty is sometimes met with, in the curling up of the posterior part of the section when transferred from spirit to oil of cloves. This can sometimes be got rid of, without spoiling the section, by careful manipulation with needles; but, if it be found impossible to obviate it in any other way, the plan may be adopted of placing each section, after it has, as usual, been stained and rinsed in water, in absolute alcohol for a few minutes, transferring it to a slide, and immediately covering it with the thin glass. Oil of cloves is then allowed to run under and clarify the specimen, which is prevented from curling up, in consequence of the pressure of the cover-glass. These precautions are not necessary in sections which have been made from paraffin and fixed to the slide.

Epithelium of the cornea.—The stratified epithelium covering the front of the cornea is well seen in the vertical sections, but the characters of the individual cells which compose it must be studied in a teased preparation. For this purpose a piece of the cornea is placed in a comparatively large quantity of $\frac{1}{8}$ per cent. bichromate of potash solution or one-third alcohol, and allowed to remain in this for a week.

Then, with the point of a scalpel or spear-headed needle, a small portion, including however the whole thickness of the epithelium, is scraped off the front, placed in a drop of distilled water on a slide, and broken up with needles as finely as possible. A piece of hair is added, and lastly the cover-glass, and the specimen is then ready for examination. The cells of the various layers will be recognised by the characteristic forms they present; those of the deepest layer being in shape like a rifle-bullet, those next above cupped to receive the rounded or conical ends of the deeper cells, and the superficial layers being more flattened as they are nearer the surface. The fine ridges and spines on many of the cells can be distinctly made out with a high power, and give a jagged contour to the cell.

To preserve the preparation permanently in glycerine it should first be stained. This is readily done by applying a drop of a very weak solution of carmalum to the edge, and allowing it to diffuse under the cover-glass; after a short time glycerine is added at the same edge, and gradually replaces the stain, the water from which evaporates meanwhile at the other borders of the cover-glass.

The substantia propria of the cornea.—The fibrous structure of the substantia propria of the cornea can readily be seen by teasing out either a fresh cornea, or one which has been macerated for a while in weak spirit or in picric acid. The lamellæ which the fibrous bundles form are apparent when an attempt is made to tear the corneal tissue, and they are well seen, cut in different directions, in the vertical sections previously made.

The corneal corpuscles are visible in the sections as mere lines, each with an enlargement in the middle, stretching across the containing cell-space, which is fusiform in section and is seldom filled by the included corpuscle. These appearances are well observed in the human cornea, and may also be made out in that of the pig and those of other animals. But although they look like mere lines in vertical section, they are, like most other connective tissue cells, in reality flattened out conformably to the surfaces of the lamellæ, and present when viewed flat great irregularities of form, and numerous branching processes with which they come in many cases into connection with one another. They are best brought to view by the gold method, and, since this also serves to show the nerves, the two structures may be studied in the same preparation.

Corpuscles and nerves of the frog's cornea.—The brain and spinal cord of a frog having been destroyed, the animal is laid on the table or held by an assistant, and the membrana nictitans of the eye is seized with forceps, and entirely removed by two or three snips with fine sharp-pointed scissors. The animal is then taken up and held in the operator's left hand, the thumb pressing upwards under the lower jaw, so that the eyes are made as prominent as possible, and the point of one of the scissor blades is inserted into the globe of the eye, just behind the insertion of the glistening, yellowish iris. By a series of snips made round the eyeball at this plane, the anterior part, with the cornea, iris and lens, is severed from the posterior, and removed to a watch-glass containing salt solution. Then, whilst the edge of the cut sclerotic is held by the one pair of forceps, with another pair the iris is seized close to the same spot, and is easily torn away from the sclerotic, bringing the lens with it. So that only the cornea, together with a narrow rim of sclerotic, is now left, and since it is floating in fluid it retains its convexo-concave form, and all crumpling of the tissue is avoided. The salt solution is now poured off, leaving, however, just enough to float the cornea in, and the watch-glass is filled up with $\frac{1}{2}$ per cent. chloride of gold solution. The cornea is allowed to remain in this a full hour; it is then removed to a beaker of water acidulated with acetic acid, and is placed in a warm place in the sunlight (see p. 120). After two days the fluid in the beaker is renewed, a teaspoonful of methylated spirit being added to prevent the growth of fungi, and in two days more the cornea may be taken out and prepared for the microscope.

It is first placed in a flat dish of distilled water, and the epithelium, which is very dark and opaque, is gently scraped off the anterior surface. This done, the sclerotic rim is cut off with scissors. It is as well to change the water at this stage, so as to get rid of the débris of epithelium. The next process consists in the separation of the corneal substance into two, three, or more thin lamellæ. With a little practice it is not at all difficult, thin as the object already seems, by holding it down at one edge with a pair of forceps and working from the same edge with another pair, to separate a very thin lamina from the concave posterior surface, consisting of the membrane of Descemet and a delicate layer of the proper substance of the cornea with its corpuscles. This posterior lamella is not only the easiest to obtain, but is also, in the frog's cornea, the most important, for it contains the closest and finest plexus of nerves. To mount it, all that is necessary is to float it on to a glass slide, to cover the preparation, and add glycerine at the edge of the cover-glass. But, since the membranous layer thus obtained has naturally the convex shape of the cornea, and it is of course desirable that it should lie flat upon the slide without creases, it is well before mounting to make three or four radial snips in it in the way shown in the adjoining cut ; these will enable it to flatten out without folding when placed on the slide and covered. Moreover it is important to examine the object with a low power previously to covering it, so that any folds of the membrane, or any foreign matter or remains of epithelium adhering to it, may be detected and removed.

Fig. 58

The remaining anterior part of the cornea may be further separated in the same way into lamellæ, which are to be mounted with the same precautions as the posterior lamella. It is not always easy to get them in quite so complete a layer, but for most purposes a small shred will, if equally thin, show nearly as much as an entire lamella.

In all these specimens the corneal corpuscles are stained of a violet colour, varying in tint according to the success of the preparation, their nuclei being left nearly unstained. The nerves are coloured almost black, the fibrils looking like fine wires running singly and in bundles, and provided along their course with numerous minute varicosities.

The corpuscles and nerves of the rabbit's cornea.—The cornea of the rabbit, or of any other mammal recently killed, is prepared with chloride of gold in the same way as that of the frog. The eyelids are first removed, care being taken in doing this not to get the hair on to the surface of the cornea. The eye is then to be made prominent, either by an assistant who holds it firmly with forceps thrust back in the orbit, so as to seize one or other of the eye muscles near their attachment, or, without an assistant, by clamp-forceps, which are inserted in like manner, and by their weight force the eyeball forward without unduly compressing it. Then the cornea, iris, and lens are removed together, after cutting round the sclerotic with scissors, and are placed in salt solution, and finally the iris and lens are removed in the same way as in the preceding preparation, and the cornea is immersed in gold solution. Since it is much thicker than the frog's cornea, it should remain in the chloride of gold—which does not penetrate very rapidly—a full hour and a half, after which it is placed in acidulated water in the light, and otherwise treated in the same way as the frog cornea. But as it is not so easy to separate the mammalian cornea into lamellæ, it is better after three or four days to place the stained cornea in gum for a few hours, when thin sections parallel to the surface may be cut with the freezing microtome, and mounted in glycerine.

Isolation of corneal corpuscles.—After the corneal corpuscles and nerves have been stained with gold in this way, they can be isolated by dissolving away the intermediate substance by caustic alkali, the action being arrested before the corpuscles and nerves, which are more resisting than the connective tissue bundles, are also destroyed. With this object a part or the whole of a gold-stained cornea—whether of frog or mammal—is placed, divested of epithelium, in a watch-glass containing a strong (20 per cent.) solution of caustic potash or soda, and this is then put into a warm chamber at 40° C. At the expiration of three-quarters of an hour the tissue, which is now quite soft and pulpy, is removed with a section-lifter, and placed in a vessel containing a large quantity of water faintly acidulated with acetic acid. Small portions may then be taken up and mounted, with or without further breaking up, in glycerine. The corpuscles are beautifully displayed, forming a continuous network by the junction of their branches; and nervous fibrils may be seen intercalated amongst the corpuscles, but never actually joined to, or continuous with, the cells.

Nerves of the rabbit's cornea.—For exhibiting the nerves of the cornea without at the same time staining the corpuscles and the epithelium, either Golgi's silver chromate method (p. 152), or Ehrlich's methyl-blue method (p. 160), or Löwit's gold chloride method (p. 159), or the following modification of the gold method recommended by Klein, may be used. The cornea of a rabbit or guinea-pig is put into ½ per cent. solution of chloride of gold for an hour and a half. It is removed from the gold into distilled water and placed in the light (without warmth) for from twenty-four to thirty hours, or until the larger nerve trunks begin to be visible near the circumference, converging towards the centre as irregular, branching lines. When the staining has arrived at this stage, and before the corneal substance generally begins to acquire a violet appearance, the cornea is removed from the water and placed in a mixture of glycerine and water (one part glycerine to two parts distilled water). After it has been in this for twenty-four hours, or longer, the corneal substance should be very little darker than before, but the nerves much more distinct, and on holding the cornea between the finger and thumb, and making sections from the anterior surface, including the epithelium and a little of the substantia propria, these, when covered in the glycerine mixture, will show not only the fine and close plexus of nerves which lies immediately underneath the epithelium, but also the far more minute network of varicose ultimate fibrils which extends between the epithelium cells almost to the anterior surface of the epithelium. If not at first sufficiently evident, these intra-epithelial nerves may generally be brought more clearly into view by placing a section for a few minutes in the strong caustic potash solution. From this it is transferred by a section-lifter to water, floated on to a slide, removed from the water, and covered, glycerine being afterwards added at the edge of the cover-glass. To prevent the latter from pressing on the softened tissue, two narrow slips of thin glass, which may be cut with a writing diamond, are to be placed one on either side, before placing the cover-glass over.

The cell-spaces of the cornea.—These are shown in two ways : by the silver method, and by the method of puncture injection.

The demonstration of the cell-spaces by the silver method may be attempted in the cornea of the frog. The animal is decapitated and the brain destroyed. The eyelid is then removed, and the eye having been made prominent by the pressure of the thumb in the manner previously recommended, the epithelium is scraped off the front of the cornea with a sharp scalpel. The cornea is then rubbed with a stick of fused nitrate of silver (lunar caustic). After five minutes the surface is thoroughly washed with a stream of distilled water from a wash-bottle. The head is now placed in spirit in the light ; in·a short time (from a few minutes to an hour), when the cornea is browned, the vessel containing it is removed from the window ·and left in a dark place for twenty-four hours. The cornea is then sliced off, placed in water, where any remaining patches of epithelium are now removed by scraping, slit in a triradiate manner, so that it may lie flat on a slide, and finally mounted in balsam.

The cornea of mammals may be prepared with silver in a similar manner, but, being generally thicker than in the frog, it is necessary to allow the caustic a longer time to penetrate, and, in the final preparation, to prepare sections parallel to the surface, instead of mounting the cornea whole.

To inject the cell-spaces of the cornea by the puncture method, the eye of the pig, or sheep, or any other animal may be used, if a human eye is not procurable for the purpose. A solution of alkanet in turpentine is the fluid which should be chosen, and the mercurial pressure apparatus (fig. 56) is used. The tube and fine steel cannula having been filled with the alkanet solution to the exclusion of air-bubbles, the cannula is inserted obliquely into the substance of the cornea, without allowing the point to pass through into the anterior chamber. The pressure is then gradually raised to about two inches of mercury, when the red fluid should begin gradually to fill the cell-spaces, and to spread through them over a considerable part of the cornea. Indeed, if the injection is long enough continued, the fluid may extend itself even beyond the corneal margin, and may penetrate into the cell-spaces in the anterior part of the sclerotic coat. The operation may with care be successfully performed without the mercurial apparatus, using merely a hypodermic syringe. But it is difficult to avoid the production of extravasation near the point of the cannula. This does not, however, always militate against the success of the experiment, for beyond the limit of the extravasation the fluid may slowly penetrate into the cell-spaces of the tissue, and this may go on even after the syringe has been removed, especially if the cornea is cut out, laid flat on a slide, and allowed to dry. As watery fluid becomes withdrawn from the cell-spaces in the process of drying the alkanet solution tends to pass in to occupy its place.

Another very successful plan of inducing the injecting fluid to pass from such an extravasation into the neighbouring cell-spaces consists in gently stroking from the extravasation towards the margin of the cornea with a smooth instrument, such as a glass rod or the ivory handle of a scalpel. But if too great pressure is exerted upon the fluid it will be found that, in place of taking the closely reticulating course which it would pursue if it merely occupied the cell-spaces, the injection tends to shoot through the tissue in straight lines which in successive planes of the corneal tissue run at right angles to one another. These lines represent the 'corneal tubes' of Bowman. Their appearance is due to the fact that the pressure exerted has been sufficient to force the injecting fluid into the interstices between the connective tissue bundles, pushing these asunder, and burrowing its way through the soft ground-substance which unites the bundles and lamellæ. And, since the cell-spaces occur in this ground-substance, the existence of a slight enlargement or fusiform swelling here and there on the tubes is accounted for.

These corneal tubes then are to be looked upon as purely artificial products, not corresponding with any pre-existing channels in the tissue (except perhaps when the fluid passes along the sheath of a nerve). They are always obtained when any fluid which is not able to penetrate into the cell-spaces is forced into the substance of the cornea. They are seen when mercury is injected by the puncture method, and this is how they were first obtained by Bowman; and if air be forced with a syringe into the tissue a similar effect is produced. In all cases the tubes cease abruptly at the corneo-sclerotic junction, where the connective tissue becomes denser, and has a less regular arrangement.

Parts at the junction of the cornea with the sclerotic.— The corneo-sclerotic junction, the ciliary muscle, and the iris are all well seen in their relations to one another in a meridional section of the part of the eye where they are situated. The section may be made from the anterior segment of an eye hardened in potassium bichromate or in 5 to 10 per cent. formol. It is not necessary to imbed this entire, but sufficient to cut out with sharp scissors under spirit a piece which includes all the parts above enumerated. The piece is cut by either the collodion or the paraffin method. Few specimens better repay the trouble of preparation than these. The cornea, sclerotic, iris, choroid, ligamentum pectinatum, canal of Schlemm, ciliary muscle, both radial and circular, and even the ora serrata and pars ciliaris of the retina, are all exhibited with the greatest clearness in a successful section, and their structure and relations may be advantageously studied.

To see the glandular depressions of the pigmented epithelium covering the choroid processes it is necessary to remove the pigment from a section by bleaching. This is effected by the action of euchlorine, chlorine water, or solution of hypochlorite of soda.

Ciliary muscle and lamina suprachoroidea.—To prepare these, the anterior part of an eye hardened in Müller's fluid is pinned down under spirit, and the cornea and sclerotic cut away at one part, when the radiating fibres of the ciliary muscle will be seen passing meridionally from their origin opposite the attachment of the iris, and forming a layer which becomes gradually thinner as it extends backwards and finally ceases in the superficial part of the choroid. A small piece of the muscle is seized with sharp forceps near its anterior border and, by carrying the instrument slowly backwards, is gradually torn away from the rest. It will be found that the shred which comes away generally spreads out posteriorly into a very thin membranous lamina, this being in fact a piece of the lamina suprachoroidea into which the superficial fibres of the ciliary muscle are inserted. A considerable length may be torn off in this way, and the piece so obtained is to be floated directly on to a slide, which is dipped for the purpose into the spirit. It must be moved with great care, so as to avoid folds or creases. The slide is quickly wiped free of spirit with the exception of that which immediately moistens the specimen, and a drop of hæmalum or picrocarmine solution is placed upon the tissue, and allowed to remain on it for ten minutes, or sufficiently long to impart its colour to the tissue. The staining solution is then poured off, and the remains of it are removed by allowing a drop or two of water to flow gently over, without disturbing the position of the membrane. Finally a cover-glass is laid on, and a drop of glycerine allowed to run in at the edge of the cover-glass.

The preparation so obtained is, if successful, a very striking one. Besides the branched pigment cells of the choroidal tissue, and a certain number of cells, similar to white blood-corpuscles, on the surface of the membrane, a number of large, round or oval nuclei are seen in the lamina, which are apparently devoid of cell-body. These are the nuclei of epithelioid cells which bound the lamina suprachoroidea externally, and serve as part of the lining of the lymphatic space which lies between this and the lamina fusca of the sclerotic. Their outlines cannot be brought to view without the aid of nitrate of silver, and the cell-bodies are too delicate and transparent to be shown by the present mode of preparation. A large number of elastic fibres are also seen on the membrane, especially at the terminations of the fibres of the ciliary muscle, where they appear to come into relation with the ends of these, an elastic fibre passing for a certain distance along each side of the muscular fibre, and seeming to serve in this way for its attachment. The involuntary fibres are particularly well shown, their nuclei being conspicuously stained by hæmateïn; many of the bundles terminate in peculiar tufts, from which the fibres radiate in all directions. It may happen that the preparation includes one of the long ciliary nerves; this, as it coursed forward to enter the ciliary muscle, having been torn away with the shred of membrane. If so, it may be followed with the microscope and its branches traced amongst the bundles of muscular fibres, forming a plexus with those of the other nerves. In tracing the branches characteristic bipolar ganglion-cells will here and there be found interpolated in the course of the nerve-fibres.

Vascular layers of choroid and membrane of Bruch.—As seen in the preparation just described, it is easy to detach the lamina suprachoroidea from the rest of the choroid.

The other three parts are more difficult to separate, and their complete isolation may require considerable time and patience. But for demonstrating their structure it is not absolutely necessary for them to be completely separated as distinct membranes; it is sufficient if, in a piece which includes all, one or other is left projecting at the edge, so as in this way to be seen distinct from the other layers. But before commencing the attempt at separation, the hexagonal pigment cells, which belong to the retina but frequently adhere to the inner surface of the choroid, must be entirely removed by gently brushing that surface with a hair-pencil. The separation and brushing are performed under fluid (spirit), and will be much facilitated by the use of a dissecting lens.

The musculature of the iris.—The circular and radiating plain muscular fibres of the iris may be demonstrated in the albino rabbit. The eye is cut in half and the anterior part placed in spirit for a day or more. Then the lens is removed, and a segment of the iris—including its whole width, from the pupillary aperture to the ciliary processes of the choroid—is cut out and placed in dilute hæmalum. When sufficiently but not too deeply stained, it is put into water for a minute or two to remove the excess of staining fluid, then passed through picric alcohol and oil of cloves, and finally mounted in balsam, with the posterior surface uppermost. The thick ring of the sphincter is easily seen in these preparations, and also the interlacing bundles of plain muscular fibres of which the dilatator is composed: they may be observed to bend round near the pupil, and take the direction of and blend with those of the sphincter. At the circumference of the iris, also, a similar bending round of the radiating fibres is observed.

Human iris.—Although, in consequence of the presence of the uveal pigment, more troublesome to prepare, it is nevertheless desirable to make a similar preparation of the human iris, for the musculature is somewhat different, the dilatator forming a uniform thin expansion, which covers the posterior surface immediately under the pigmented epithelium, and not distinct radially arranged interlacing bundles with intervening meshes as in the rabbit. The specimen may be made from an eye that has been in Müller's fluid or bichromate of potash (3 per cent.), preferably not in spirit. A piece is cut out as before, and is treated in a similar way, except that before being stained the pigment is brushed completely off the posterior surface with a stiff camel-hair brush. This must be done under fluid, and of course very carefully so as to avoid tearing the tissue; during the operation the iris is examined now and then with a low power, to determine when all the pigment is removed. It is difficult to prevent some of the pigment granules from still sticking to the surface, but, as they tend for the most part to adhere along the lines of junction of the fibre-cells, their presence does not spoil the object of the preparation, for the dilatator fibres are if anything better displayed.

A method of getting rid of the remains of the pigment is to treat the piece of iris, after brushing, with chlorine water or solution of hypochlorite of soda until completely bleached: then wash, stain, and mount as above.

THE RETINA

Sections of the retina.—It will be well to study the general arrangement of the several layers of the retina by means of sections, before its constituent elements are observed isolated. The best method of hardening is by a mixture of three parts Müller's fluid and one part of 1 per cent. osmic acid, the process being completed by spirit. Corrosive sublimate may also be used. If possible to obtain it perfectly fresh and healthy, the human retina should always be taken; if not so obtainable, that of the pig is preferable to the retina of most other common mammals.. For the study of the central fovea, the human eye or that of a monkey is necessary.

To harden the retina rather small pieces are cut out from

different parts, or, what is better, the posterior part of the bulb is kept whole, so that the membrane remains supported by the outer coat. This posterior part should be turned inside out and the vitreous removed, and it is then dropped into a relatively large quantity of the hardening fluid, and allowed to remain in it for a week, the fluid being stirred at intervals and changed on the second day. After a week it is transferred first for two or three hours to water, then for twenty-four hours to 50 per cent. alcohol, and then to strong spirit. In another day or two it is ready for the preparation of sections. Small pieces from different regions may be taken, but as they are all treated in the same way they may be described as one.

The staining of the tissue is first effected by placing the piece for twenty-four hours in alcoholic logwood (Kleinenberg's) or alcoholic fuschin. From this it is transferred to spirit, then embedded in paraffin. It is better so to place it in the embedding tray that the sections shall be both vertical and meridional, since, made in this way, they will take the general course of the fibres of the optic nerve. The sections cannot be made too thin, but they should be complete, that is to say, they should include all the layers of the membrane. They are cut in series, fixed with water (p. 37) and mounted in balsam. But if hardened with corrosive sublimate or corrosive and picric they are best stained on the slide, and for this any of the ordinary methods, and especially methyl blue or toluidin blue and eosin (p. 23), may be employed.

Preparation by the Golgi method.—As Cajal has shown, no method shows the relations of the retinal elements to one another better than the method of Golgi. The best way to employ it is as follows :—Place small pieces of fresh retina from the ox, dog, or pig—preferably pieces which are rolled up upon themselves—in a relatively large quantity of the osmium-bichromate mixture (one part 1 per cent. osmic to three parts Müller) : keep them in the dark for from one to five days : each day take a piece, and after rinsing it in 0·75 p.c. nitrate of silver solution transfer it to a quantity of the same solution containing a trace of formic acid (1 drop to 100 c.c.). After forty-eight hours in the silver solution wash with distilled water, transfer to 95 per cent. alcohol for an hour or two ; place in collodion for a few minutes, fix on a metal holder with the collodion, and immerse in 95 per cent. alcohol ; then cut the sections with a sliding microtome, pass them rapidly through absolute alcohol and bergamot oil or xylol, and mount, without a cover-glass, in xylol balsam. A large number of sections should be mounted in this way and searched. Some will show only Müllerian fibres ; others rod and cone elements ; others the inner granules, and others again the nerve elements of the molecular layer : all, of course, stained black by the reduced metal.

Ehrlich's methylene-blue method (p. 160) may also be employed for retina, with great advantage, and after fixation by Bethe's fluid followed by alcoholic solution of platinic chloride (1 in 300), sections may be made by the paraffin method.

Isolation of the retinal elements.—Various processes are employed for macerating portions of the retina in order to obtain its elements, either in a completely isolated condition or still partially connected with one another. It will be best, in the first place, to attempt the separation with a piece of retina which has been in 1 or 2 per cent. osmic acid for six or eight hours. It must of course be put in perfectly fresh, and after the time mentioned it is placed in water for twenty-four hours. It is then allowed to macerate for a few days in a mixture of equal parts of glycerine, alcohol, and water, after which a minute portion is to be carefully broken up with fine needles in a drop of weak glycerine, and, a piece of fine hair having been added, the cover-glass is superposed and tapped to effect the separation and isolation of the elements. The piece of retina may be kept for years in the alcohol glycerine and water mixture, and the longer it is left the easier becomes the isolation.

Other portions of fresh retina may be placed in 10 per

cent. chloral hydrate solution for two or three days. The portions so macerated are to be teased out in a drop of the same solution, the usual expedient being adopted of obviating the pressure of the cover-glass by a hair.

In the chloral hydrate preparation, the rod- and cone-elements are well preserved, as are also most of the other structures. The external segments of the rods, which even in the osmic preparation tend for the most part to become altered, may by this method be frequently seen almost unchanged, and the transverse striation, which indicates their discoid formation, is often well marked.

The study of the retina cannot be considered complete until the elements have been examined in the fresh, unaltered condition. A small piece, taken from an eye still warm from the animal, should accordingly be broken up as rapidly and finely as possible in a little vitreous humour. A modification of this very simple method consists in allowing a small piece to macerate for two or three days in weakly iodised serum (serum or amniotic fluid in which iodine has been shaken up) before attempting the dissociation, which can then be more readily effected.

The **hexagonal pigment** of the retina may be seen in most of the teased preparations above described. In eyes that have been hardened in Müller's fluid the layer often separates in flakes of varying size, and nothing is simpler than to remove such a piece with a section-lifter, and mount it in glycerine, so as to exhibit the pavemented appearance which the cells present.

Silver preparation.—On the innermost surface of the retina, also, a mosaic-like appearance can be demonstrated by the aid of nitrate of silver, but it is much more irregular, and does not depend upon the presence of epithelium cells, but upon the flattened-out ends of the Müllerian fibres. To show this appearance, a fresh eye is cut in half transversely, and the vitreous is shelled out from the posterior half, which is turned inside out; this is then rinsed in distilled water and transferred to nitrate of silver solution ($\frac{1}{2}$ per cent.). After a minute in this it is again rinsed in distilled water, and exposed in water to the light. When the retinal surface is browned the eye is removed, a piece of the retina is cut out under water, floated on to a slide with the inner, brown surface uppermost, and the water is drained off, the preparation allowed to dry, and then mounted in balsam. Or it may be mounted, without drying, in glycerine, care being taken not to let the cover-glass press upon the specimen.

The retina in the lower Vertebrata.—The structure of the outer segments of the rods, which is difficult to make out in mammals, can be seen easily, even with an ordinary high power, in the retina of amphibia. With this object the eye is removed from a recently killed frog, cut across, and a small portion of the retina is quickly broken up in vitreous humour. A piece of hair having been added, the preparation is covered and examined. Almost everywhere the field of view is strewn with large, clear, rod-shaped structures, some straight, but many of them bent and curved in different directions, and exhibiting a distinct transverse striation, or even a tendency to split up into a number of superimposed discs, this tendency increasing as the preparation is longer made. Some of them have what looks like a small appendage jointed on at one end; this is the comparatively small inner segment of the rod. The cones are also small in comparison, and on that account may at first be missed; they are distinguished by the possession at the apex of the inner segment of a small, bright, fatty globule, often of a yellow colour. Most likely a portion of the hexagonal pigmented epithelium will have come away with the rest of the retina, and in consequence of the rupture of some of the cells the preparation will be strewn with pigment granules which, like all minute granules suspended in fluid, exhibit very strikingly the Brownian molecular movement. Some of the pigment cells may be observed intact, either isolated or in patches. If seen in profile, it may be noticed that near one surface (the outer) the cell is almost entirely free from black pigment, while from the other fine streamers of the cell-protoplasm, dotted with pigment granules, extend. In their natural position these pass between and amongst the outer segments of the rods in the eyes of an animal which has been kept in the light, but are confined to the region of the cell-nuclei in eyes which have been kept in the dark, or which are taken from a curarised animal (Kühne). The cones also vary in length under these conditions, being greatly shortened in the eye of an animal kept in the light (Engelmann). This difference is best shown in sections of retinæ which have been hardened in corrosive sublimate.

The retina of a bird, of a reptile (tortoise), and of a fish may be teased out fresh in vitreous humour, in the same way as that of the frog. The chief points of interest in these preparations are the ellipsoid or lenticular bodies in the inner segments of the rods (bird and amphibian) and cones (bird, reptile, and amphibian), the bright, fatty globules of different colours in the inner segments of the cones in the tortoise and bird, and the twin or double cones, especially large in the fish's retina. The various other points in which the retina in these animals differs from that of mammals, may be studied by employing the same methods of preparation as or the mammalian retina.

THE LENS AND VITREOUS HUMOUR

The lens fibres.—The following will be found the best mode of isolating the fibres of the lens, as well as for showing their arrangement. Take the fresh eye of any animal—that of the ox or sheep for example—and cut it across into an anterior and a posterior half. Place the anterior part, having removed what remains of the vitreous humour, in one-third alcohol. Then scratch through the posterior capsule, which is readily ruptured and curls away from the lens proper. This can easily be shelled out, and is left in the fluid, the remainder of the eye being rejected. The lens is allowed to remain in the alcohol for two or three days, being merely turned over once or twice. It will be found that its substance tends both to separate along the radiating lines which mark the planes of junction of the ends of the fibres, and also to peel into concentric lamellæ like the coats of an onion ; and, if a piece of one of these lamellæ is taken up with the forceps, it will tear in the direction of the fibres from one of the planes of junction of the anterior surface to the corresponding plane of the posterior. The fibres can be readily separated with needles. For this purpose portions should be taken both from the superficial and from the more central parts of the lens. In

THE EYE 279

many of the superficial fibres a round or elongated nucleus may be detected at one part, and since the nuclei of adjacent fibres are met with in about the same region, when a number of fibres are seen together the nuclei lie in an irregular row. The riband-like shape of the fibres may be made out at parts where they are turned over so as to be seen edgeways.

Sections of the lens.—For cutting sections of the lens it is best to harden it in formaldehyde (5 to 10 per cent. formol). The whole anterior half of an eye should be put in this, the cornea having been partly removed so as to enable the fluid to get freely to the front as well as to the back of the lens, but the capsule is not to be ruptured. After two or three days, the preparation may be soaked in gum, and sections may be made by the freezing process. They are to be mounted in glycerine. The lens must not be put in spirit to complete the hardening, for strong spirit renders the tissue, especially the central parts, quite hard and horny in consistence, and the outlines of the fibres become obliterated.

The epithelium of the lens-capsule.—This may have been seen in the antero-posterior section as a single row of nucleated cells, lying immediately behind the anterior part of the capsule. To show it on the flat it is to be stained with nitrate of silver. With this object a lens still enclosed in its capsule is removed from a fresh eye, and, after having been rinsed in distilled water, transferred for five minutes to $\frac{1}{2}$ per cent. nitrate of silver solution. It is then again washed with distilled water, and placed in the light in weak spirit (equal parts spirit and water). When brown it is removed from the light, and placed in 75 per cent. spirit. After twenty-four hours it is hard enough to allow tangential sections to be made from the anterior surface, to include the capsule, the epithelium, and the parts of the lens substance immediately subjacent to this. The sections are mounted either in glycerine or in balsam with the brown surface uppermost : and through the elastic capsule, which is not distinctively stained, the outlines of the epithelium-cells are clearly seen. At places the silver solution may have penetrated to the superficial lens fibres, and will be found to have stained the cementing substance between them.

The zonule of Zinn and the hyaloid membrane of the vitreous humour.—Take the anterior half of the eye (preserved in spirit or formol) of an albino rabbit, and having pinned it, the cornea downwards, on a loaded cork, and removed the remains of the vitreous humour, gently seize the lens with fine forceps, and draw it away from the iris. In doing this it will drag with it the suspensory ligament, the zonule of Zinn, and the part of the hyaloid membrane continuous with this, so that the separated lens appears girdled by a delicate, somewhat crumpled-looking, membranous zone, closely adherent at its inner border to the equator of the lens, and bounded outwardly by a ragged margin—the torn edge of the hyaloid. Cut out with fine scissors a segment of this zone, including its whole breadth, and with a section-lifter transfer the piece so removed to hæmalum solution or some other dye. When sufficiently stained—and it stains very readily—transfer it to a dish of water, and from this float it on to a slide, avoiding all creases except of course the natural ones of the zonule. It may then be covered, and the water in which it is mounted replaced by glycerine. Or, instead of placing it in the water, it may be transferred from the logwood to spirit, and then passed through oil of cloves and mounted in balsam. These preparations exhibit well the folds and striations of the zonule, and the rounded corpuscles, like white blood-corpuscles, which are dotted here and there over the surface of the hyaloid.

THE BLOOD-VESSELS OF THE EYE

For the demonstration of the blood-vessels the head of an albino rabbit should be injected, a cannula being placed in each carotid, and the two cannulas connected to the arms of a Y-shaped tube, the stem of which is brought into communication by an indiarubber tube with the injection bottle.

After the blood has been driven out of the vessels, before the flow of injection fluid, the neck of the animal, just below the place where the cannulas are inserted, is surrounded by a loop of wire, which is drawn as tightly as possible to prevent the escape of the injection ; and the pressure is then raised to about four inches of mercury and kept there for some minutes, so as to make certain that all the blood-vessels shall be completely filled. The whole is then allowed to stand and become cold, that the gelatine may set, after which the eyes are to be carefully excised, and placed in 3 p.c. bichromate, free incisions being made in the sclerotic. After two or three weeks they are tranferred to spirit, first 50 per cent., then 75 per cent., then strong. When they have been in this last a day or two the following parts may be prepared :—

The conjunctival vessels, and the subconjunctival vessels of the sclerotic.—By making with a razor held in the hand and wetted with spirit a tangential section from the region of the corneo-sclerotic junction, and after passing the piece so obtained through oil of cloves, mounting it in balsam with the outer surface uppermost, the distribution of the vessels is exhibited, both in the conjunctiva and, by focussing more deeply, those in the sclerotic at and near the margin of the cornea. Another plan consists in cutting away a small piece, including the whole thickness of both cornea and sclerotic, and mounting in a similar way. The thickness and irregularity of the piece so obtained is a disadvantage, but, on the other hand, the canal of Schlemm and the other venous sinuses may be observed, if the injection has been a successful one, by focussing still lower than for the looped vessels of the sclerotic.

Vessels of the choroid and iris.—One of the two injected eyes is to be divided by an antero-posterior cut with the razor into a right and a left half. One of the two halves, the one which does not include the attachment of the optic nerve, is first taken, and the vitreous, retina, and lens removed, so as to clear the inner surface of the choroid and iris. The last-named are next to be separated as one piece of membrane from the sclerotic. The piece so obtained is then to be again divided into two, by another antero-posterior cut with the scissors, and the resulting halves are to be mounted, after passing as usual through oil of cloves, in balsam, the one with the inner and the other with the outer surface uppermost. Each includes, of course, the fourth part of the choroid coat with some of the ciliary processes, and a piece of the iris ; and with a low power the course and disposition of the blood-vessels in these parts can be readily followed. Besides these comprehensive preparations, separate ones may be made from the other half of the eye of a portion of the iris (this is rendered more instructive by lightly staining it), and one or two of the ciliary processes snipped off with sharp scissors, and mounted so as to be seen in profile.

The vessels of the retina.—If the other injected eye be cut into an anterior and a posterior half, and the posterior part is examined after removal of the vitreous humour, the blood-vessels will be seen spreading out from the centre of the colliculus of the optic nerve. To exhibit their finer distribution in the retina, a piece is mounted flat in balsam without previous staining, while, to show the extent of their distribution in the retinal layers, vertical sections, which need not be very thin, may be made from a piece embedded in paraffin in the ordinary way, and similarly mounted, without staining, in balsam.

CHAPTER XXII

THE EAR

THE only parts of the ear which require special directions for their preparation are the semicircular canals and the cochlea.

The semicircular canals.—To study the structure of the membranous semicircular canals, those of the cartilaginous fishes, *e.g.* the skate, are chosen. The skull, which can be readily cut with a scalpel or strong pair of scissors, is opened quite anteriorly, where it is occupied merely by a quantity of cerebro-spinal fluid, and the opening is extended backwards by removing the roof bit by bit, until the whole of the upper surface of the brain is exposed. Two thick cartilaginous masses will be seen, one on either side, near the posterior part ; the large auditory nerves pass through a foramen in each into their interior. These masses enclose the membranous labyrinth, consisting in these animals of utricle, saccule, and semicircular canals, all of large size, and contained in corresponding cavities and canals, in the substance of the cartilage, but of which no trace can at present be made out. If, however, horizontal slices are made with a scalpel, one of the canals will soon be exposed, and this can then be followed in both directions by cutting the cartilage away so as to expose the included membranous canal in its whole length. It will be found to lead at either end into a large membranous bag— the utricle—with which the two other canals also communicate and from which they can be traced in the same manner. Besides the utricle, there is another smaller membranous bag

—the saccule—and both contain a white, pasty, cretaceous, otolithic mass, which lies over the part to which the nerve proceeds. Near one of the attachments of each semicircular canal to the utricle is its dilated part, or ampulla, and a branch of the auditory nerve may be seen proceeding to each of these, and terminating abruptly in a forked thickening, which indents the membranous wall and lies transversely to the axis of the ampulla.

The three ampullæ, and the adjacent portions of the semicircular canals, are now to be removed from the cavities containing them, and are to be placed, one in Flemming's or Hermann's solution, one in corrosive sublimate or corrosive sublimate and picric acid (p. 18), and the third in 1 per cent. osmic acid. The one in Flemming's or Hermann's solution is transferred to weak spirit after three days, and in twenty-four hours more to strong spirit. After another day or two it may be stained with Kleinenberg's logwood or some other bulk stain, and then imbedded in paraffin. Sections are to be made both of the semicircular canal proper and of the ampulla, opposite to and including the entrance of the nerve. The one in corrosive sublimate is transferred for a few hours first to iodised spirit, and then to absolute alcohol; after a day or two in this it may be embedded and cut. The sections can be stained after being fixed with water to a slide. The third ampulla, which was placed in osmic acid, is transferred in an hour or two to water, and after another hour to a mixture of equal parts of glycerine, alcohol, and water. After a day or two in this it is broken up in dilute glycerine, and examined with a high power, with the view of observing the two kinds of epithelium cells—columnar and spindle-shaped—which occur, and the stiff, hair-like projections which are attached to the former. The ampullæ may also be prepared by the Golgi method to show the nerve-terminations.

Sections of the cochlea.—On account of the thinness of its osseous parietes, the ease with which it may be obtained separate from the surrounding bone, and its comparatively large number of spiral turns, the cochlea of the guinea-pig offers far greater facilities for study, and especially for the preparation of sections, than that of any other readily available animal. The following is the mode of finding and procuring it :—In the recently killed animal the aperture of the mouth is prolonged backwards on either side, by cutting through the cheeks and temporal muscles with strong scissors. The lower jaw is then seized and forcibly torn away from the rest of the head, so that the base of the skull is exposed. Here will be seen on either side, just behind the fossa for the articulation of the condyle of the jaw, a large white bony projection— the tympanic bulla. This is not yet to be opened, but the cartilaginous external auditory meatus is first cut through, and with the aid of bone-forceps or strong scissors, the bulla in question, together with the petrous bone to which it is attached, is separated from the rest of the skull. In a young animal this can be very readily effected, simply by inserting a strong blunt instrument into the base of the skull just in front of the bulla, and, using it as a lever, raising the bone and forcing it away from its attachments. The bones of either side being thus removed, the adhering soft parts are cleared away and the bulla is broken open at its most prominent part. On now looking into the cavity there will be noticed on one side the delicate tympanic membrane stretching over the end of the external meatus, with the handle of the malleus attached to it, and on the opposite wall a well-marked conical projection, which there is no difficulty in recognising as the cochlea; indeed, its bony wall is so thin that it is possible to count the number of turns (four) which it presents. By cutting the bulla round with strong scissors, the two parts—one including the tympanic membrane, and the other the cochlea— are separated from one another, and the membrane part may at once be dropped into 5 p.c. formol and put aside to be subsequently stained and mounted. From the other part as much as possible of the substance of the petrous bone is snipped away, bit by bit, from around the base of the cochlea with scissors or bone forceps, but great care should be taken in approaching the cochlea itself, as this is very readily split. When the surrounding bone has been in this way removed, the cochleas are dropped into Flemming's or Hermann's fluid or corrosive sublimate solution to which is added an excess of crystals of picric acid. When the bone is completely softened, a process which is much facilitated by frequent disturbance of the fluid, the cochleas are placed in 50 per cent. alcohol and transferred gradually to stronger spirit, which is repeatedly changed. When the excess of picric acid is entirely removed they may be embedded and cut in a plane running through the axis of the modiolus. The cochlea may have been stained in bulk before being embedded, or the individual sections may be stained after being fixed on the slide. Only a few complete axial sections can be obtained from each cochlea.

Teased preparations of the cochlea.—Successful sections will show the general position and relations of the rods and other parts, and to a certain extent the individual elements. But only a profile view can in this way be obtained, and since the minute structure of the elements composing the organs of Corti can only be properly seen when isolated, it is necessary to prepare other cochleas with this object in view. Another animal is accordingly sacrificed, and the cochleas removed as before. One is placed in a 1 or 2 per cent. solution of osmic acid : the other in a 0·05 per cent. solution of chromic acid ; but, before dropping them into their respective fluids, the bony wall must be cut through here and there with a scalpel, so that the fluid shall at once penetrate to the interior of the turns. The cochlea in osmic acid is removed after three or four hours and placed in water. After two or three days both cochleas may be further prepared in the following way :—

The uppermost turn is broken or snipped off with scissors, placed in a drop of water on a slide, the shell of bone which forms the cupola and outer wall removed, and the piece of lamina spiralis examined with a low power (without covering the preparation) in order to learn to recognise the structures which lie on it. The glass slide is then removed to the dissecting microscope, and with very fine needles the lamina spiralis is separated from the columella, which is then rejected. Next all the parts on the lamina, but especially the row of rods of Corti, to which the hair-cells as a rule cling, are broken up finely, but at the same time slowly and carefully, the preparation being examined now and again with the highest power which it is safe to use without a cover-glass. One of the chief difficulties is apt to arise from portions of the tissue sticking to the needles : if this is the case, pieces of glass rod drawn out to a fine point may be substituted. When the more important parts have been broken up pretty completely, any thick pieces of tissue unimportant to the present observation, such as bits of bone, or periosteum, bundles of medullated nerve fibres, &c., should be picked out, and then a cover-glass laid on and the preparation examined. To preserve either preparation permanently glycerine may be allowed to diffuse in at the edge of the cover-glass ; but the chromic specimen should first be treated with a drop of picro-carmine solution, so that the elements are somewhat stained.

In this way a number of specimens may be obtained from each cochlea—proceeding from above, down, and preparing turn after turn ; and careful sketches should be made of the different structures met with, and their arrangement with regard to one another. It will be found that the osmic preparations serve best for showing the lamina reticularis and the lamina basilaris, and the chromic preparations for the hair-cells and the membrana tectoria ; the other structures are almost equally well seen in both kinds of preparations. The large fat-droplets in some of the epithelium cells of the uppermost turn are peculiar to the guinea-pig, as is also the arched projection—seen in the sections—at the part where these cells are found. The fat-drops are stained black in the osmic preparation.

THE OLFACTORY ORGAN

Sections of the olfactory mucous membrane.—Very small pieces of the upper turbinate bones, or from the upper (olfactory) region of the septum nasi, from the dog or rabbit, are placed, one piece in a quantity of 0·2 per cent. chromic acid solution or in Flemming's solution, a second in 0·2 per cent. bichromate of potash, or 0·05 per cent. chromic acid, and a third in 1 per cent. osmic acid for three or four hours and then in water. The one in chromic acid or Flemming may remain a week, when it is transferred to weak spirit and then gradually to strong spirit. After a day or two in this, vertical sections are prepared from it.

Isolation of olfactory epithelium cells.—The other pieces are examined after forty-eight hours' maceration, small pieces being teased out so as to isolate the epithelium cells (both columnar and spindle-shaped), and if possible, especially in the osmic preparations, to study the connection of their branching lower ends with subjacent structures. These preparations can be preserved with glycerine, the chromate ones being stained with logwood.

Teased preparations should also be made of the olfactory mucous membrane of the frog or newt. Having cut off the head of the animal, and slit up the nostrils with fine scissors, place the head in a quantity of 0·2 per cent. solution of bichromate of potash or 0·05 per cent. chromic acid. After two days, preparations of the epithelium from both the anterior and posterior part of the passage may be made. The cells are obtained with the greatest ease by scraping the mucous surface with the point of a scalpel, and shaking out the material in a drop of water on a slide. A piece of hair is added, and the preparation covered and examined. In the portion obtained from near the anterior nares ordinary columnar ciliated epithelium cells will be seen. In that from the true olfactory part the cells, although many of them are of a columnar form, are destitute of cilia, and in addition to the columnar elements, spindle-shaped (olfactory) cells are met with which are provided with a bunch of stiff-looking, hair-like processes, resembling the similar appendages of the auditory epithelium.

The connection of the olfactory cells with the olfactory nerve-fibres may be studied in preparations from fœtal animals made by the method of Golgi.

THE GUSTATORY ORGANS

Taste-buds.—For studying the taste-buds, the foliated papillæ which are found on either side of the base of the rabbit's tongue are used. To obtain them the tongue is cut

Fig. 59

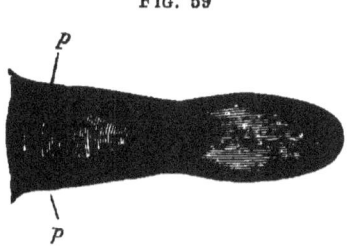

Tongue of rabbit, seen from above
p, p, Papillæ foliatæ

out entire from the recently-killed animal; when the little oval patches marked with transverse ridges may readily be found (fig. 59, *p, p*). They are snipped off with curved scissors, and one is dropped into a mixture of one part of 1 per cent. osmic acid and three parts of Müller's fluid, and the other is placed in 1 per cent. osmic acid. After three days the one in osmic-bichromate is divided, and one half is prepared with silver nitrate by Golgi's method, whilst the other half is placed in strong spirit. Both halves are eventually cut into sections, vertical to the surface of the mucous membrane, and across the direction of the ridges.

U

The piece that was placed in osmic acid is transferred to water after three or four hours, and may then be used as follows :—In the first place, two or three sections are to be obtained like those made from the other piece ; they can be made by the freezing method, and are to be mounted in glycerine. One such section is to be placed in a drop of water on a slide, and an attempt made with needles, under the dissecting microscope, to separate some of the taste-buds from the surrounding epithelium. For this purpose the needles must be very fine, sharp, and clean, and the lens used as high as is consistent with convenience of manipulation. When one or more taste-buds have been thus separated, the rest of the section is removed, and the isolated buds are broken up as completely as possible into their constituent cells. The specimen may then be covered, and a drop of glycerine allowed to diffuse in under the edge of the cover ; after which an examination of the preparation may be made, at first with the ordinary high power, and afterwards with an immersion objective.

INDEX

ABSORPTION of fat, 226
Acetic acid, use in histology, 16
Acid fuchsin, 21
Adipose tissue, 112
— — development of, 113
Alcohol, use in histology, 17
Alkanet for injecting lymphatics, 192
Ammonia bichromate, for hardening nervous tissue, 252
Aniline, 21
— blue-black, for staining nervous tissue, 253
— blue (Nicholson's No. 1), for staining parietal cells, 191
— dyes, 21
Appliances for microscopic work, 51
Areolar tissue, 96
— — action of acetic acid on, 99
— — corpuscles, 97
— — — staining of, 99
— — constrictions on fibre bundles, 100
— — elastic fibres, 97, 106
— — — — stained with magenta, 100
— — fibres of, 96
— — interstitial injection of gelatine into, 101
— — preparation by method of localised œdema, 100
— — prepared with nitrate of silver, 104
Arytenoid cartilage of ox, 121
Asphalte solution for injecting lymphatics, 192
Attraction particles, Henneguey's method of showing, 88
Axis-cylinders in spinal cord, 151

BERLIN-BLUE, mode of preparing, 179
— — reduction of, in tissues, 184

Berlin-blue, solution for injecting lymphatics, 192
Bethe, method of fixing methylene-blue preparations, 160
Bichromate of potash as a fixing solution, 16
— — ammonia, 16
Bismarck-brown. *See* Vesuvin
Bladder. *See* Urinary bladder
— of frog, 135
Blood, human, action of reagents on, 58
— — — acetic acid, 60
— — — alkalies, 61
— — — chloroform, 61
— — — tannic acid, 61
— — — water, 59
— — mode of obtaining, 48
— — coverglass preparation of, 80
— — on warm stage, 55
— — — effect of superheating, 57
Blood-corpuscles, human, red, appearance varies with distance of objective, 50
— — — effect of reagents, 58
— — — salt upon, 51
— — — warmth upon, 57
— — — water, 59
— — — observation of, 49
— — — structure of, 58
— — — method of fixing and preserving, 79
— — mode of counting, 63, 66
— — human, white, 51
— — — amœboid movements of, 57
— — — development of, 113, 132
— — of frog or newt, 66
— — — — action of electric shocks on, 71
— — — — action of reagents on, 74
— — — — — iodine, 74
— — — — feeding of white corpuscles, 69

u 2

Blood corpuscles of frog, migration of white corpuscles, 70
— — — — influence of warmth on white corpuscles, 71
— — — — Stirling's method of preserving, 81
— — of newt, action of boracic acid, 75
— — — — — — carbonic acid, 76
— — — — — steam in fixing, 80
Blood crystals, 77
— platelets, 52
— vessels, development of, 113
— — injection of, 176
— — larger, epithelial lining of, shown by nitrate of silver, 152
— — — — elastic layers of, 164
— — — — fenestrated membrane of, 164
— — — muscular tissue of, 164
— — — mode of hardening, 165
— — — sections of, 165
— — — sub-epithelial layer of, 163
— — smaller, epithelial cells of, 166
— — — muscular structure and nuclei, 167
Bone, corpuscles in lacunæ, 125
— developing, 129
— hard, grinding section of, 123
— — precautions in mounting, 124
— softened in hydrochloric acid, 125
— — — chromic acid, 125
— — — picric and other acids, 127
— — lamellæ and Sharpey's fibres, 127
Boxes for microscopic specimens, 14
Brownian movement, 55

CABINET for keeping specimens in, 14
Cajal, modification of Golgi's silver chromate method, 152
— duplicate process, 153
— method of fixing methylene-blue, 160
Camera lucida, for delineating objects, 41
Canada balsam, solution in xylol, 25
Cannulas for injecting, mode of preparing, 182
Carminate of ammonia, 20
Carmine gelatine injection, 177
— solution for staining tissues, 20
Cartilage, articular, in fresh state, 114
— — vertical and tangential sections of, 116
Cartilage, cell spaces of, 117
Cartilage-cells, action of water on, 114

Cartilage-cells, preservation of, 115
— — stained by chloride of gold, 118
Cartilage, costal, 120
— matrix, cell territories of, 120
— — stained by logwood, 121
— transition between hyaline and yellow, 121
Caton's apparatus for studying circulation in fish, 171
Cedar-wood oil, for use with immersion objectives, 5
Celloidin. See Collodion
Cell spaces of connective tissue, 104
Central nervous system, modes of preparing, 252–256
Central tendon of diaphragm prepared with nitrate of silver, 189
Centrosomes. See Attraction-particles
Cerebellum. See Central nervous system
Cerebrum. See Central nervous system
Chloral hydrate for preparing retina, 275
Chloroform for dissolving paraffin, 31
Choroid coat of eye, 271
— — — blood-vessels of, 281
— — — — lamina suprachoroidea, 270
— — — layers of, 271
Chromic acid as a fixing solution, 15
— — — — dissociant, 16
Ciliary motion, action of reagents on, 91
— — — — carbonic acid on, 92
— — — — chloroform on, 93
— — — — warmth on, 91
— — — — weak alkalies, 92
Ciliary muscle, 270
Circulation in omentum of guinea-pig, 172
— — frog's web, 169
— — lung of toad, 173
— — mesentery of toad, 171
— — tails of tadpoles and fishes, 170
— — tongue of toad, 174
Clearing fluids, 24
Cochlea, mode of procuring, 284
— precautions for embedding and cutting, 285
— softening of osseous parietes, 285
— teased preparations of, 285
Cohnheim's gold method, 118
Collodion method of preparing sections, 29
— — — fixing sections, 87

Condenser, bull's-eye, use of, 3
— — — used as dissecting lens, 7
— substage, 3
Conjunctiva, blood-vessels of, 281
Connective tissue corpuscles in areolar tissue, 97
— — — — tongue of toad, 175
— — cell-spaces of, 104
Cornea, cell-spaces of, 267
— — — — mode of injecting, 268
— epithelium of, 261
— mode of hardening, 260
— precautions to avoid curling up of sections, 261
— substantia propria of, 262
— of frog, corpuscles and nerves of, 263
— — rabbit, corpuscles and nerves of, 265
— — — isolation of corpuscles, 265
— — — nerves of, 266
Corneal tubes of Bowman, 269
Corneo-sclerotic junction, 269
Corrosive sublimate, as a fixative, 18
Cover-glass holder, 9
Cover-glasses, mode of averting pressure of, 68
— mode of cleaning, 8
— — — fixing, 26
— — — — by paraffin, 25
— selection of, 8
— forceps for, 9
— preparations, mode of making, 80

DAMMAR varnish, 26
Decalcification methods, 127
Decolourisation, 22
Delineation of microscopic objects, 40
Demidesiccation method, 96
Dentinal sheaths, 214
Dextrine used in freezing method, 28
Diaphragm of microscope, use of, 3
— — — — for observing fresh tissues, 85
— central tendon of, 189
— lymphatics of, 189, 196
Directions for work, 39
Dyes, 18

EAR, 283
Egg-white for fixing sections to slide, 37
Ehrlich, methylene-blue method of, 160
— — — — applied to retina, 274

Ehrlich-Biondi triple stain, 23
Elastic fibres in areolar tissue, 97, 106
— — transverse section of, 107
— networks of serous membrane, 107
— — — artery, 164
— tissue, 107
Electricity, mode of applying, 71
Embedding in celloidin, 29
— — paraffin, 31
— by gum method, 202
Endocardium, 206
— fibres of Purkinje in, 206
End-plates of mammals, 157
— — — lizard, 158
Eosin, as a stain, 21
— — — combined with hæmateïn, 22
— — — methyl-blue, 23
— — — toluidin blue, 23
Epidermis, cells of horny layer shown by potash, 84
Epithelioid-cells, 104
— — covering tendon, 112
Epithelium, ciliated, from frog's mouth, 89
— — — gills of mussel, 90
— — study of separated cells, 94
— columnar, 84
— olfactory, 288
— scaly, of mouth, 82
— — deeper layers of, 83
Epithelium-cells, fibres in, 88
Erectile tissue, mode of hardening, 247
Eye, blood-vessels of, 280
— general mode of preparing, 257
— of pig as substitute for human eye, 258
Eyelids, sections of, 258
Eye-piece of microscope, 4

FAT ABSORPTION, 226
Fat-cells, 112
— — membrane of, 113
— — development of, 113
Fibrin in blood, 52
Fibro-cartilage, yellow, of epiglottis, 122
— — transition to hyaline, 121
— — white, 122
Fibrous tissue, 107
Field-glass, 4
Flemming, method of staining karyokinetic figures, 87
Flemming's fluid, 16
Forceps, 11

Formaldehyde, 17
Formol, as a fixative, 17
Freezing method, 28
Fuchsin. *See* Magenta
Fuchsin acid, 21

GANGLIA, sections of, 152
— Ganglion cells, 152
Gas, carbonic acid, mode of applying to a preparation, 76
— chamber, 68
Gastric glands, 222
— — cells of isolated, 223
Gelatine injecting fluid, 177
— — mode of preserving, 185
Gentian-violet, 21
Glycerine for mounting, 24
— jelly for mounting, 24
Glycogen, its presence in white blood-corpuscles, 74
— its presence in liver-cells, 236
Goblet cells, 86
Gold chloride, methods of staining with, 118, 159
Gold size, for fixing cover-glass, 25
Golgi, silver chromate method of, 152
— — — — applied to study of bile-ducts, 235
— — — — — central nervous system, 252
— — — — — retina, 274
Granules in blood, 52
— — Osler's observations upon, 52
Gullet. *See* Œsophagus
Gum, used in freezing method of preparing sections, 28
— — for embedding, 202

HÆMALUM, 19
— Hæmatein, 19
— combined with eosin, 22
Hæmatoxylin, 19
— acid, 19
— Delafield's, 19
— Kleinenberg's, 20
Hæmin crystals, 78
Hæmoglobin crystals, 77
Hairs, 201
— development of, 203
Hamilton's method of staining degenerated nerve-fibres, 256
Hardy and Westbrook, on granules of white-corpuscles, 80
Haversian canals of bone, 124
— fringes of synovial membrane, 198
Hayem's fluid, 64

Heart, muscular substance of, 205
— lymphatics of, 207
— blood-vessels of, 207
Heidenhain's bulk-stain, 20
Hepatic cells, 236
Hermann, method for showing karyo-kinetic figures, 88
Hermann's fluid, 16
Hoggan's rings, 189
Holmgren, apparatus of, for studying circulation in lung, 173

IMMERSION objectives, mode of using, 5
Inflammation changes in mesentery of toad, 172
— — — tongue of toad, 176
Injected parts, mode of preparing, 183
Injection apparatus, 179
— of an entire animal, 181
— — blood-vessels, 176
— — — mode of killing an animal for, 181, 233
— — lymphatics, 192–96
— mass, mode of preparing, 177
— with nitrate of silver, 185
Injections fluid in the cold, 184
Instruments required for microscopic preparation, 7
Intercellular substance, shown by silver nitrate, 89
Intestine, large, 230
— small, blood-vessels of, 227
— — mode of hardening, 225
— — nerves of, 227–30
Iodised serum, 276
Iris, blood-vessels of, 281
— diaphragm of microscope, 3
— muscular tissue of, 272
— sections of, 269
Iron, micro-chemical test for, 24
— its presence in liver-cells, 236

KANTHACK and Hardy, on fixing granules of white corpuscles, 80
Karyokinesis, method of staining, 21, 88
— — fixing, 87
Kidney, blood-vessels of, 243
— examination in fresh condition, 244
— mode of hardening, 242
— tubules, isolation of, by Ludwig's method, 243
— — basement membrane of, stained with nitrate of silver, 244

Klein, method of showing nerves of cornea, 256
Kleinenberg's hæmatoxylin, 19
Koch's (von), method, 213

LABIA, mode of preparing, 247
Lachrymal gland, 259
Lamina fusca, 260
— suprachoroidea, 270
Larynx, 179
Lens, isolation of fibres, 278
— sections of, 279
— suspensory ligament of, 280
— capsule, epithelium of, 279
Lilienfeld and Monti, micro-chemical test for phosphorus, 24
Liver, blood-vessels of, 232
— lymphatics of, 235
— mode of hardening, 231
— — injecting bile-ducts, 234
Logwood solution for staining tissues, 19
— — Kleinenberg's, 20
Löwit, gold chloride method of, 159
Ludwig's apparatus for injecting lymphatics, 193
Lung, injected with paraffin, 212
— epithelium of air-cells, 210
— injection of blood-vessels, 211
— mode of hardening, 209
— of toad, circulation in, 173
Lymphatic glands, 237
Lymphatics, injection of, 192
— of diaphragm, natural injection of, 196
— — tendon, mode of injecting, 193
— larger, 192
— smaller, in omentum and central tendon, 186–90

MACALLUM, micro-chemical test for iron, 24
Magenta, 21
— double staining by, 22
— solution for fresh tissues, 22
Magnifying power of microscope, estimation of, 47
Mammary glands, 251
Mann's fluid for hardening, 18
— — — staining, 23
Marchi's solution, 254
Marrow, 181
— red, on warm stage, 132
— — giant cells of, 133
Mayer's carmalum, 20
— hæmalum, 19

Measurement of an object under the microscope, 44
Methyl-blue, 21
— — and eosin, 23
Methyl-green, 21
Methyl-violet, 21
Methylene-blue, 21
— — method of Ehrlich, 160
Mesentery, stained with nitrate of silver, 188
— circulation in, 171
Micrometers, 44, 45
Microphotography, 44
Microscope, binocular, use of, 3
— eye-piece of, 4
— for dissection, 7
— magnifying power of, 46
— parts of, 1
— powers of, 4
— selection of, by student, 6
Microtome, freezing, 29
— inclined plane, 80
— tripod, 33
— rocking, 34
— Minot's, 36
Migration of white corpuscles from veins, 176
Modelling, 41
Moist chamber made with putty, 61
Mould for paraffin, 32
Mouth, mucous membrane of, 213
Mucous glands of tongue, 217
Müller's fluid, composition of, 16
— — used for eye, 257
Muscle, blood-vessels of, 145
— ending in tendon in mouse's tail, 144
— — — — frog, 144
— examined by polarised light, 142
— transverse section of, 137
— of insects, 139
— — mode of production of the transverse striæ, 140
— — prepared by Rollett's method, 141
— — — — — studying contraction of, 140, 141
— involuntary, mode of isolating cells of, 134
— — — showing nuclei, 134
— — prepared with nitrate of silver, 135
— voluntary, action of acetic acid, 137
— — demonstration of sarcolemma in, 137
— — mammalian, 136, 137
— — isolation of fibres, 138

Muscle, voluntary, separation of into discs and muscle-columns, 138
Myeloplaxes, 133

NAIL, cells of, separated by potash, 84
— embedded by gum-method, 202
— sections of, 202
Needles, mounted, 11
— spear-headed, 11
Nerve cells of ganglia, 152
— — — spinal cord, 150
Nerve fibres, degenerating, 150
— — medullated, 146
— — — treated with osmic acid, 147
— — — stained with nitrate of silver, 148
— — non-medullated, 147
Nerve trunk, perineurium of, shown by silver method, 148
— — sections of, 149
— — structure of, 148
Nerves, motor, ending of, 157
— — methods of staining, 158
Nervous plexuses of intestine, 227–230
Nervous system, central, modes of preparing, 252–256
Neuroglia-cells, 150
Nissl, method of, 254
Nitromolybdate of ammonia for fixing methylene-blue, 21, 160
Nose-piece, 5
Nuël, fibres in epithelium cells, 88

OBJECT-GLASS, or objective, 4
Objectives, immersion, 5
— apochromatic, 5
Ocular of microscope, 4
Œsophagus, 220
— blood-vessels of, 220
Oil of bergamot, for clearing, 25
— — turpentine, 25
— — cloves, for clarifying sections, 25
Olfactory mucous membrane, 288
Oliver, G., method for estimating number of blood-corpuscles, 66
Omentum, prepared with nitrate of silver, 186
— circulation in, 172
Optical section, 169
Orange G, 21
Organs of taste, 289
Osler on blood-platelets, 52

Osmic acid, as a fixing solution, 16
— — colours fatty substances black, 16
— — for nerve, 147
— — for retina, 273–275
Ossification, intracartilaginous, 129
— intramembranous, 129
— of lower jaw, 129
Ovary, 248
Ovum, mode of obtaining, 248

PACINI'S fluid for blood-corpuscles, 79
Pacinian corpuscles from cat's mesentery, 154
— — sections of, 157
— — treatment with chromic acid, 156
— — — — nitrate of silver, 156
— — — — osmic acid, 155
Pal's method. See Weigert-Pal
Palate, 218
Pancreas, 236
Paraffin, for fixing coverglass, 24
— method of preparing sections, 31
— mould for, 32
— for embedding, 31
Pericardium, 205
Perineurium, 148
Phloroglucin in decalcifying bone, 127
Phosphorus, micro-chemical test for, 24
Pia mater, vessels of, 168
Picric acid, as a fixative, 17
Picrocarmine, 20
Pigment, hexagonal, of retina, 276
— of ciliary processes, method of bleaching, 270
Pipettes, mode of making, 11
— measuring, for blood enumeration, 67
Pituitary body, 237
Pleura, 208
Polarisation apparatus, 5
Polarising microscope, 142
Prostate gland, 247
Pulmonary vessels, injection of, 211
Purkinje, fibres of, 206

RANVIER on Sharpey's fibres, 128
Ranvier's demidesiccation method, 96, 98
— methods of preparing areolar tissue, 102

Ranvier's method of showing tendon-cells, 110
— — — membrane of fat-cell, 113
Reagents in common use, 14
Recklinghausen's silver method, 103
Remak, fibres of, 147
Reticular tissue, 238
Retina, blood-vessels of, 282
— fibres of Müller, 275, 276
— fresh in vitreous humour, 276
— in iodised serum, 276
— isolation of elements, 275
— methods of hardening and cutting sections, 273–275
— of bird, reptile, and fish, 278
— of frog, 277
Rollett, gold method of, applied to muscle, 141
Rubin, S. *See* Acid fuchsin

SAFRANIN, 21
Salamander tadpoles for karyokinetic figures, 87
Salivary corpuscles, 83
— glands, 218
Salt solution, normal, 15
Sanderson and Stricker, on circulation in omentum, 173
Schaffer, on decalcifying methods, 128
Sclerotic coat of eye, 259
— — — — blood-vessels of, 281
— — — — lamina fusca, 260
Scissors, 11
Scrotum, 247
Section-lifters, 11
Sections, modes of preparing, 26
— — — keeping in series, 34
— methods of fixing upon slide, 36
— — — mounting, 39
— optical, 169
Semicircular canals, mode of finding and preparing, 283, 284
Seminal vesicles, 247
Seminiferous tubules, epithelioid cells of, 250
— — isolation of, 250
Serous membranes, 186
Sharpey, fibres of, 127
Shellac for fixing sections, 37
Sherrington, fluid for diluting blood for enumeration, 64
Silver chromate method. *See* Golgi
— nitrate, method of staining epithelium with, 89
— — — — connective tissue, 104
Skin, 199
— blood-vessels of, 201

Skin, double staining of, 200
— preparation of, 199
— nerves of, 203
— lymphatics of, 203
Slides for microscopic purposes, 8
— — counting blood-corpuscles, 64
Solutions for examination of fresh tissues, 14
— — fixing the tissues, 15
— — staining histological objects, 18
Spinal cord, 253
— — isolation of cells of, 150
Spleen, 239
— demonstration of retiform tissue of, 240
— injection of, 239
Stage of microscope, 1
— mechanical, 5
Staining fluids, 18
— double and triple, 22
Steam for fixing amœboid corpuscles, 80
Stirling's method of staining and preserving frog's blood-corpuscles, 81
— — — showing medullated nerves in trachea, 212
Stomach, 221
— blood-vessels of, 223
— glands of, 222, 223
— horizontal sections of, 223
— lymphatics of, 224
Stomata in lymphatic septum of frog, 191
— central tendon of diaphragm, 190
Suprarenal capsule, 5, 241
Synovial membranes, 197
— — blood-vessels of, 198
— — Haversian fringes of, 198
— — preparation of by silver method, 106, 197
Syringe, hypodermic, for interstitial injection, 101
— — for injecting lymphatics, 195
— condensing, for injection, 181

TACTILE corpuscles, 201
Tarsal cartilage of eyelid, 259
Taste-buds in papillæ foliatæ of rabbit, 289
— — — circumvallatæ, 217
Teeth, in situ, 215
— development of, 216
— dentinal sheaths of, 214
— sections obtained by grinding, 213
— — — v. Koch's method, 213
— sections of, softened, 214
— soft tissues of, 215

x

Tendon of mouse's tail, 108
— — — — action of acetic acid on, 109
— — — — cell-spaces of, 111
— transverse section, 110
— cells, 109
Tenon, capsule of, 259
Testis, 249
— lymphatics of, 249
— tubules of, 250
Thionin, 21
Thoma, decalcification method, 127
Thymus gland, 237
Thyroid body, 237
Toluidin-blue, 21
— — and eosin, 23
Tongue, 217
— blood-vessels of, 218
— of toad, 174
Tonsils, 218
Trachea, 212
— epithelium of, 94
Trays for microscopic specimens, 14
Tunica vaginalis, 251

URETERS, 244
— epithelium of, 245
Urinary bladder, 245
— — of frog, 135
Uriniferous tubules. *See* Kidney
Uterus, 248

VAGINA, 247
— Vaginal synovial membranes, 197
Vesiculæ seminales, 247
Vesuvin, 21
Vitreous humour, hyaloid membrane of, 280
Villi, structure of, 225

WALLERIAN degeneration, 255
— Warming apparatus, with gas regulator, for chloride of gold preparations, 119
Warm-box, 58
Warm stage, simple, 53
— — — mode of estimating temperature, 54
— — with gas regulator, 55
Weigert's method for neuroglia, 255
Weigert-Pal method for central nervous system, 253
Weil. *See* Koch
White of egg for fixing sections, 37

XYLOL-BALSAM, 25
— Xylol for clearing, 25
— — dissolving paraffin, 33, 37

ZONULE of Zinn, 280

www.ingramcontent.com/pod-product-compliance
Lightning Source LLC
Chambersburg PA
CBHW021954220426
43663CB00007B/816